Commemorations
and Memorials
Exploring the Human Face
of Anatomy

Commemorations
and Memorials
Exploring the Human Face
of Anatomy

editors

Goran Štrkalj
Macquarie University, Australia

Nalini Pather
University of New South Wales, Australia

World Scientific

NEW JERSEY · LONDON · SINGAPORE · BEIJING · SHANGHAI · HONG KONG · TAIPEI · CHENNAI · TOKYO

Published by

World Scientific Publishing Co. Pte. Ltd.
5 Toh Tuck Link, Singapore 596224
USA office: 27 Warren Street, Suite 401-402, Hackensack, NJ 07601
UK office: 57 Shelton Street, Covent Garden, London WC2H 9HE

Library of Congress Cataloging-in-Publication Data
Names: Štrkalj, Goran, editor. | Pather, Nalini, editor.
Title: Commemorations and memorials : exploring the human face of anatomy /
    edited by Goran Štrkalj, Nalini Pather.
Description: Hackensack, NJ : World Scientific, 2017. |
    Includes bibliographical references and index.
Identifiers: LCCN 2016059281 | ISBN 9789813143142 (hardcover : alk. paper)
Subjects: | MESH: Tissue and Organ Procurement | Gift Giving |
    Anatomy--education | Cadaver | Ceremonial Behavior
Classification: LCC RB35 | NLM WO 660 | DDC 611.0022/8--dc23
LC record available at https://lccn.loc.gov/2016059281

British Library Cataloguing-in-Publication Data
A catalogue record for this book is available from the British Library.

Typeset by Stallion Press
Email: enquiries@stallionpress.com

We dedicate this book to donors of the past, present and future; for their noble act of giving the ultimate gift to medical research and education — their own bodies after death.

The Editors

PREFACE

We are more than just our bodies. Over the past 100 years, the content of human anatomy has remained relatively stagnant in comparison to the body of knowledge which comprises how anatomy is taught and assimilated by learners and adapted to health care settings. Despite a significant reduction in contact hours for didactics, an increased emphasis on electronic media, and even with the widespread curriculum stretch to introduce and incorporate new imaging techniques in the classroom, the majority of gross anatomy programs still, with intentionality, dedicate time to acknowledge the altruism of body donation.

As an anatomy teacher, I have attended many student-run ceremonies meant to honor body donors and express gratitude to the families of the deceased. Some ceremonies were small and informal; one comes to mind in which the students simply placed a rose on the dissection table and dedicated some time for personal reflection before leaving the anatomy laboratory after their final examination. Others were large memorial ceremonies; I can recall one in which students handed roses and wild flower seed packets to several hundred family members who were invited to attend a celebration of thanksgiving with student-lead musical performances, poetry, and a candle-lighting while names of the donors were read aloud. Bearing witness, I must say that the moment in which the eyes of the anatomy student and the family member of the donor meet in the simple exchange of a flower is poignant. There is anticipation is on both sides. The family of the donor seems eager to see those who have benefited from the gift of their love ones. In a way, the family's attendance is an act of seeking assurance in the decision to donate; families want to see with their own eyes the new generation of health care providers for whom their loved one was a pivotal teacher. Students, on the other hand, are eager to catch a glimpse of their body donor in living family members — some shred of human connection to anchor the many wonderings the students have harbored about the person whose body became known to them so intimately. The donors themselves, in giving their bodies to education, become medical students' most memorable teachers. Ceremonies that facilitate this intersect provide positive and healing closure for both, emotional and cathartic relief. Additionally, when students recognize and express gratitude for donors and their families, the act undoubtably enhances the ever-emerging sense of professional responsibility.

This book entitled *Commemorations and Memorials: Exploring the Human Face of Anatomy*, edited by the esteemed anatomy professors and medical educators Drs. Goran Štrkalj and Nalini Pather, is a carefully selected collection of papers which together present the humanitarian aspects of anatomy in review of various commemorative services. This is an invaluable resource for any anatomy teacher, in any culture, and in all parts of the world to learn from and reflect on the variety of ceremonies that exist to celebrate and honor the gifts of body donors. In addition to detailed descriptions of ceremonies around the world (including: Brazil, China, Australia, New Zealand, South Africa and United States) the reader can also find in-depth reflection from the past dark era of anatomy during World War II, and thoughtful analysis on the current state of affairs in donation programs outside of academia. Bioethical issues which arise in these ceremonies are also considered.

After reading this book it will be clear to the reader that thanksgiving ceremonies have an important role in the curricula of health professions programs using human cadavers in their anatomy courses. Students clearly benefit from this positive and healing closure after performing human dissection — an intellectually and emotionally intense experience.

A few years ago when she was still alive, the late Mary A. Feeley, revered Mayo Clinic chaplain, concluded the thanksgiving ceremony with a dedication not from Scripture, but from one of her favorite country music songs by Wynona Judd, "Every light begins with darkness; every flower was once a seed." She explained that we all should celebrate darkness and light, seed and flower, and the wonder of both earth and sun. In her words, "the donor's death has offered a glimpse into the wonder and power of the human spirit." It is no surprise that she decided to donate her body to the students to whom she ministered some years later, extending that very same wonder she intimated beyond her own life.

With this book we commemorate the ways in which we are touched by acts of remembrance — the very acts vital to honoring that which makes us human, that which makes us more than just our bodies.

Wojciech Pawlina, M.D.
Professor and Chair
Department of Anatomy
Mayo Clinic
Rochester, MN,
February, 2017

CONTENTS

INTRODUCTION

Goran Štrkalj* and Nalini Pather⁺

*Department of Chiropractic, Macquarie University, Sydney, Australia
⁺School of Medical Sciences, UNSW Australia, Sydney, Australia

For a considerable number of students studying human anatomy today, the educational journey of exploring the human body might not begin with a tutorial or a dissection session, or even with a customary introductory lecture. In an increasing number of institutions, it commences with a commemoration of the body donors — the people whose dead bodies, now embalmed cadavers, are to be dissected and studied during the anatomy course. Whereas these ceremonies are most often conducted in teaching spaces, they can also take place around specially created memorial objects (monuments) and in memorial spaces (memorial halls and parks). While both memorial ceremonies and objects/spaces are not new in the discipline of anatomy, they were in the past presented and carried out only sporadically and in an isolated manner. Since the mid-20th century, they have become more common and have been adopted by anatomy departments around the world, with a significantly increased frequency in the last few decades.

Up until recently there were relatively few publications dealing with the ways anatomists honor body donors. It is therefore not surprising that the terminology used for anatomy commemorations is highly diverse: "commemoration", "memorial service", "service of thanksgiving", "dedication service", "convocation of thanks", and "service of gratitude" among others (Tschernig and Pabst, 2001; Hildebrandt, 2010; Pawlina *et al.*, 2011; Jones *et al.*, 2014). In some cases, the difference in terminology is only sematic in nature whereas in others, it differentiates the modes in which these commemorations are organized and carried out. We suggest (with no intention to be prescriptive in proposing terms for very complex entities) that, for the sake of convenience, all these phrases should be subsumed under the umbrella term "commemoration" to signify all ceremonies and activities held to pay respect to donors, help families and friends of the donors in the grieving process to have some closure, and to demonstrate to the community or its representatives that the bodies (cadavers) in anatomy laboratories are treated not only in accordance with legal requirements, but also with moral integrity. Some of these ceremonies may include information about how the bodies are used in the laboratory and their importance to research and education. We also suggest that the term "memorial" should be used for the objects and spaces that serve to honor and remember those whose bodies are used in research and in the course of teaching and learning anatomy. Commemorations (occasions and services) and memorials (objects and spaces) are often closely connected in serving the same or similar purposes.

This book, the first of its kind, examines the emerging phenomena of anatomy commemorations and memorials. The recent trends in anatomy commemorations and memorials can be understood and contextualized by examining the history and current status of the discipline of anatomy, especially in regard to anatomy teaching and learning.

Anatomy is one of the oldest scholarly disciplines in human history (Persaud *et al.*, 2014), its roots extending to prehistoric times when basic knowledge of the body structures was necessary for efficient hunting and butchering and for the early healing practices. In more recent times, research in anatomy provided the backbone for the development of modern life sciences and medicine. Similarly, in the realm of education, anatomy has been one of the key subjects within the curricula of many science and medical disciplines. During its long history, anatomy education has endured numerous transformations including the most recent phase of accelerated change in educational theory, rapid advancements in the sciences and medicine, and a surge in learning technologies. Indeed, in recent years anatomy education has been intensively reviewed and investigated (Louw *et al.*, 2009; Pabst, 2009; Drake *et al.*, 2009, 2013; Sugand *et al.*, 2010; Drake *et al.*, 2014; Chan and Pawlina, 2015). As noted recently "over the last few years there have been enormous numbers of papers published on modifications of the anatomical curriculum, while there are hardly any studies on similar topics regarding teaching of chemistry, biology or physiology" (Pabst, 2009). So numerous are the innovations in anatomy teaching that two American anatomists could confidently declare at the beginning of the 21st century: "The old anatomy is dead. Long live the new anatomy" (Reidenberg and Laitman, 2002: 81). Furthermore, anatomy is now seen as a discipline that can provide students with much more than knowledge of the structure of the human body. Modern anatomy curricula embed a range of capabilities including professionalism, research skills and clinical competencies in its teaching and learning practice (Netterstrom and Kayser, 2008; Bockers *et al.*, 2010; Kerby *et al.*, 2011).

One of the key elements in this recent transformation of anatomy has been a change in attitude towards human cadavers used in teaching, and in their treatment while in the anatomy laboratory. This new approach, in turn, has been rooted in a more reflective attitude of anatomists, and in attempts to suffuse anatomy education with humanistic values. The discipline of anatomy has become more humane, inspired by constant reference to the humanities, which form the axis of reflective practice. The change of attitudes towards cadavers and human tissue in modern anatomy is so dramatic that it is sometimes referred to as cultural (Dyer and Thorndike, 2000) or a paradigm change (Talarico, 2013). Although at varying rates in different countries and teaching traditions, anatomy is now eminently emerging as a deeply transformed discipline, as a new, humanistic anatomy (Štrkalj, 2014).

The history of the use of cadavers in anatomy is complex and controversial (Persaud *et al.*, 2014). Cadavers played important role in anatomy research and teaching albeit in different ways, influenced through the years by varying social and cultural contexts as well as diverse philosophies of changing medical and scientific paradigms. Dissection of human bodies flourished only briefly during Antiquity, in the Hellenistic

Alexandria. Taking advantage of the permissive cultural atmosphere of the times, Herophilus and Erasistratus carried out numerous dissections of human cadavers (and possibly also vivisections) as a powerful tool in the study and teaching of the structure and function of the human body. Dissections were revived in the late Medieval times with the first recorded public dissection being carried out by Mondino de Liuzzi at the University of Bologna at the beginning of the 14th century. Until the 18th century, dissection was both a public and teaching event, usually held in winter to prevent rapid decomposition of the body. Somewhat simplified, it can be stated that in Medieval and early Renaissance periods the main purpose of cadaveric dissection was to confirm what was written in the classical medical texts, primarily those of Galen. It was Andreas Vesalius who dramatically changed this practice and started to investigate the human body using dissection as his primary tool. This enabled him to significantly extend the knowledge of human anatomy and correct many mistakes present in the works of the classical authors. In spite of considerable opposition, the Vesalian approach prevailed and dissections became a prime research and teaching tool in anatomy. This practice continued until the present day, although the role of dissection in education is questioned and hotly debated. It appears, however, that the majority of experts maintain that cadavers, either through student dissection or inspection and study of prosected bodies, still constitute the key element of anatomy education (Pather, 2014).

Cadavers for dissection were obtained in various ways, usually, but not always, sanctioned and regulated, sometimes illegally, often in morally dubious ways and against the scorn of the general public. The bodies used in dissections were traditionally those of the disenfranchised, lower strata of society. At first, the cadavers used in anatomy were those of executed criminals, however, throughout history cadavers provided to anatomy schools in this way were not sufficient in numbers to satisfy the need of developing medical education. In 1831, for example, only 11 bodies were legally available (decided upon by English judges as part of the death sentence for those found guilty of murder) for dissection in London which had more than 900 anatomy students (Lucas, 2013). Anatomists, consequently, often engaged in illegal activities including stealing cadavers from mortuaries and graveyards.

In the late 18th and early 19th centuries, organized gangs of grave robbers satisfied the ever-increasing demand for cadavers by anatomists in Britain and the USA. The most extreme instance of this hideous enterprise occurred when some of these gangs realized that they could get bodies faster and more easily by killing people. The most famous of these cases was that of the notorious Burke and Hare pair who killed no less than 16 vulnerable (poor and homeless) individuals and sold their corpses to the Anatomy School of the renowned Edinburgh anatomist Robert Knox (Rosner, 2011). Body snatching and similar disturbing activities would start to disappear with the introduction of the 1832 Anatomy Act in Britain. This Act, following the practice of some of the other European countries such as France, allowed anatomist to appropriate the unclaimed bodies from hospitals, asylums and shelters for the poor. The Anatomy Act thus rendered dissection as a punishment for poverty (Richardson, 2001). Historians recount a number of strategies employed to increase the number of unclaimed bodies available for dissection, such as the hospital's failure to inform the immediate family of a patient's death (Richardson, 2001, 2008).

While the demand for unclaimed bodies grew, body donations for anatomical dissection were generally rare. Most of the donors were educated and enlightened individuals, such as philosopher Jeremy Bentham in England, who understood the importance of body bequests, and were not constrained by prevailing cultural norms. In the Western world, donations became more common in the second half of the 20th century as access to unclaimed bodies diminished and alternate ways of obtaining cadavers were needed. This was not due to ethical issues related to appropriation of unclaimed cadavers but rather to the changing social milieu (increased standards of living and access to social support for funeral costs) and cultural values (decreased need for elaborate funeral practices). Today, whereas the laws of many countries allow only donated bodies to be used in anatomy laboratories, there remains a considerable number of countries that still rely mainly on unclaimed cadavers (including those imported from other countries) and even bodies of executed criminals for their anatomy instruction (Anyanwu *et al.*, 2011).

Once in the anatomy laboratories, cadavers were generally treated rather dispassionately. To avoid the stress attached to working with the dead bodies and to help medical students develop clinical detachment, students were taught to be unemotional and treat cadavers as objects. It is not a surprise therefore that historians describe the 19th century anatomists and their students as "a brutish group" (Macdonald, 2005). This type of behavior and practices in medical schools would, in some countries, periodically get out of hand and even cause public riots.

This does not mean that students were not encouraged to respect the dead. Students were taught to "focus on the science of medicine and ignore the art — the human face — of medicine" (Dickinson *et al.*, 1997). The strategy of "detached concern" was the key coping mechanism utilized and unquestioned for years (Keniston, 1967). One was expected to detach in order to be able to carry out emotionally charged tasks (such as dissecting dead human bodies) while at the same time preserve concern and respect. Balancing the two, however, was no easy task (one even wonders if "detached concern" is an oxymoron) and it appeared that the detachment often overpowered the concern. In a process that has heightened in many institutions in recent decades, the attitudes towards cadavers has been changing profoundly. Students, rather than being detached, form a professional medical relationship with a cadaver, and often the family of a donor. One of the key components in this newly established relationship are commemorations and memorials which became crucial in paying respect and connecting with donors, their families and community.

The ceremonies are carried out, and memorials constructed, in different ways, depending on the social and cultural factors characteristic of the community within which they originate (Tschernig and Pabst, 2001; Park *et al.*, 2011; Jones *et al.*, 2014; Subasinghe and Jones, 2015). Some ceremonies are organized by faculty with no or only marginal organizational participation of students. Others, on the other hand, are organized by both faculty and students, and in many cases by students alone. Some include relatives and friends of the body donors, some not. Some are organized once a year and some several times a year. Donors are kept anonymous in some ceremonies whereas in the others their identity is revealed with consent of the donor and family. Commemorations might include religious representatives or be completely secular. They often include artistic performances, reading of reflective texts and might include a variety of ritualistic performances (such as prayers and pledges) and symbolic objects (such as flowers and candles). Furthermore, as one would expect nowadays, electronic resources, internet and social media often play important role in communicating and performing these ceremonies.

Notwithstanding the form in which they are carried out commemorations represent elaborate social occasions which include contemplative activities that have many positive outcomes while addressing different stakeholders. Indeed, services and memorials constitute part of a greater, complex social network within which modern anatomy education is being carried out (Bolt, 2012). Commemorations and memorials provide an opportunity for the academic community to reciprocate the generosity of the donors and their relatives for their gift — their own bodies or that of their loved ones after death. They enable academics and students to show respect and gratitude to donors, and also add to students' confidence that their study using bodies is well justified and does not transgress any of the cultural or social norms.

Anatomy ceremonies also provide powerful strategies that help decrease students' anxiety and uneasiness in handling the dead (Martyn *et al.*, 2013). Furthermore, it has been noted that anatomy education could be used efficiently in the first stages of developing professionalism (Lachman and Pawlina, 2006; Slotnick and Hilton, 2006). The ceremonies and their preparations, in conjunction with other activities, could not only be used efficiently in this process in medicine and allied health professions, but also in other disciplines in which practitioners will be dealing with human remains, such as biological anthropology and forensic science.

Family and friends of the donors are another close party participating in many of these ceremonies. For those who were close to the deceased donors, commemorations offer an opportunity to vow the final farewell to the loved ones and help in the grieving process. Memorials built in some institutions are places where these families and friends can return to, for the remembrance and to perform ceremonies related to the deceased. The sheer number of people attending the commemorations and visiting memorials indicates their importance as social functions. For example, anatomists at the Radboud University Nijmegen Medical Centre in the Netherlands, were somewhat surprised with the high number of relatives and friends of the body donors — on average 10 persons per donor — who appeared at the unveiling ceremony of the body donor monument (Kooloos *et al.*, 2010; Bolt, 2012).

For potential donors, their families and community as a whole, commemorations and memorials provide a reassurance that the bodies in anatomy laboratories are treated with respect and that body donation is of great importance in education. Commemorations and memorials, of course, cannot by themselves guarantee that nothing untowardly has, or will not happen in the laboratory as, sadly, abuse can occur even in the most regulated and organized places. However, they are a clear sign that modern anatomists are attempting to do their best. They also demonstrate that anatomy departments are not closed and secretive institutions as they often used to be in the past. The departments are rather structured as forward-looking scientific and/or medical units open to public gaze and criticism, welcoming comments and creative input. Modern anatomy is therefore not about death, it is about life — celebrating the past while improving the present and building the future. The commemoration ceremonies and monuments thus forge a powerful bond of trust between the scientific/medical community and the general public as anatomists demonstrate that nothing distrustful happens in the anatomy laboratories and highlight the importance of cadavers in research and education.

In 2009, one of us (GŠ) was engaged in developing a new anatomy program and preparing resources for the operating of a soon-to-open new dissection laboratory. For the first time in the history of this relatively young university, cadavers were to arrive to the laboratory, and an important part of the new program was to be a memorial ceremony for the body donors. In preparation for this, and in attempt to devise the most appropriate mode of memorial service, a literature survey was carried out. Somewhat surprisingly, a relatively small number of papers were found. The search was further complicated by the before-mentioned terminological heterogeneity. In addition, information about the ceremonies was often meagre, a short reference or a paragraph in articles dealing with issues such as body donation or aspects of anatomy education. Whereas in the following few years the number of papers on the topic increased significantly, their total number is still relatively small, particularly those providing a detailed account of the ceremonies and their analyzis. At the same time, however, a comparatively large number of items on memorial services are available in popular writing and media, particularly on the internet and in various social media. It was not surprising to find, for example, that a paper on memorial ceremonies in Korea has almost as many references to the internet material as to the papers published in academic journals (Park *et al.*, 2011). The prevalence of information in the media and in the public domain is another clear sign of the importance of anatomy commemorations and public interest in them.

An appraisal of the literature demonstrated that in some countries surveys of anatomy departments were carried out to establish if, and how, memorial ceremonies were performed. In Germany, of 36 departments (all responded), 81% held the ceremonies (Tschernig and Pabst, 2001), whereas in the USA out of 84 departments (68.2% of those to which questionnaires were sent) 95.5% carried out the ceremonies (Jones *et al.*, 2014). The manner in which these are organized and carried out is diverse. Data from most other countries is not available but one can hypothesize that in many the ceremonies are beginning to emerge (Kooloos, 2010; Oxley da Rocha *et al.*, 2013; Zhang *et al.*, 2014; Subasinghe and Jones, 2015).

At the same time, it would appear that a considerable number of anatomists have become acutely aware of the need to humanize anatomy. Indeed, several professional bodies and informal groups of academics have issued noteworthy statements and recommendation with regards to the use of cadavers and, more broadly, on anatomical collections (McHanwell *et al.*, 2008; Champney, 2011; American Association of Anatomists, 2009; Participants, delegates and supporters of the International Conference on 'Cultures of Anatomical Collections', 2012). Some of these documents include statements on memorial services. The Trans-European Research Group, for example, issued a set of recommendations for the Good Practice for the Donation of Human Bodies and Tissues for Anatomical Examination. Among the recommendations is the following (which was later also absorbed into the recommendations of the International Federation of Associations of Anatomists): "Medical schools or anatomy departments should be encouraged to hold Services of Thanksgiving or Commemoration for those who have donated their bodies for medical education and research to which can be invited, relatives of the deceased, staff and students" (McHanwell *et al.*, 2008).

There is, of course, no ideal commemoration that could work in all circumstances. Building up on the idea of a community of practice, the purpose of this book is to enhance the exchange of ideas and sharing of experiences. As two German anatomists noted about anatomy services of thanksgiving, "anatomists should discuss this topic within their own departments and with the students and learn from each others [sic] experiences, not only within their own country but also from the situation in other countries with different historical and cultural backgrounds" (Tschernig and Pabst, 2001).

Indeed, it appears that approaches developed within one culture might have successful application outside that original cultural milieu. For example, a pilot study research carried out at the University of Michigan suggested that medical students at this institution prefer regarding cadavers in the anatomy laboratories as "teachers" as they are treated in Medical Schools in Thailand (Winkelmann and Güldner, 2004) and Taiwan (Lin *et al.*, 2009), rather than "first patients", a practice more common in the Western countries (Bohl *et al.*, 2011). In favor of this "Eastern approach" is also the fact that anatomy is taught in the early stages of medical education whereas the relationship with patients seems to be still new and foreign to the students. The authors of the Michigan study proposed an innovative approach that could combine the two traditions where "anatomy programs could introduce the donor first as a teacher and later transition into viewing donor as a patient" (Bohl *et al.*, 2011).

In this book, we do not intend to be prescriptive or to offer a definite word on the subject. In this introduction, we present only a broad delineation (offered in the beginning) rather than strict definition of the key terms. We understand the concepts of "commemoration" and "memorial" in anatomy programs as emerging. We see these as concepts which will continue to evolve and ramify within different cultural and educational contexts. The book is a collection of texts that utilize a variety of approaches to focus on ceremonies and memorials in anatomy programs and includes both case studies and broader theoretical discourses. The book starts with a chapter written by Gareth Jones who investigates the ethical dimensions of commemorations in anatomy. In Chapter 2, John

McLean looks at commemorations and memorials from a philosophical perspective and argues that they can help embed the concept of human flourishing into medical practice. In Chapter 3, Thomas Champney discusses the lack of memorial services in for-profit and non-profit will body companies in the USA. In Chapter 4, Sabine Hildebrandt looks at one of the darkest periods in the history of anatomy, that of the Third Reich, and reflects on the ways to commemorate the victims of Nazi prosecution whose bodies were used in anatomy teaching and research. Carl Stephan, Jodi Caple, Andrew Veprek, Emma Sievwright, Vaughan Kippers, Steve Moss and Wesley Fisk suggest an ethical method to acquire human skeletons for scientific study and propose a mode of memorializing the persons whose skeletal remains were used in anatomy schools without their consent, in Chapter 5. These five papers are followed by six chapters presenting a variety of ways in which commemorations are carried out in different cultures and academic traditions: Ernest Talarico in the USA, Natasha Flack, Kathryn McClea and Helen Nicholson in New Zealand, Beverley Kramer and Graham Louw in South Africa, Jiong Ding and Luqing Zhang in China, Andréa Oxley Da Rocha, Deivis De Campos, João Antônio Bonatto-Costa, Júlia Pedron, Maria Paula Oliveira de Moraes in Brazil, Nalini Pather and Ken Ashwell in Australia. In Chapter 12, Yousef AbouHashem, Benjamin Brown and Goran Štrkalj investigate the presentation of anatomy commemorations on YouTube. The book ends with few concluding remarks from the editors of this volume which outline the possible areas for future studies.

Marginalized and abused in the past, the human body dissected in anatomy laboratory has been regaining centrality in the process of education in recent years and has been transformed from a "teaching resource" into a "teacher". While commemorations and memorials provide an opportunity to pay respect to the donors and express gratitude for their gift, they are only a part of the process in which anatomy education reinvents itself in an effort to become a more humane discipline. Ernest Talarico, in his contribution to this book, succinctly expressed this approach to education when he noted that "the teaching philosophy itself is a commemoration and tribute to the donor." It is hoped that this book will contribute towards this new philosophy becoming a norm.

REFERENCES

American Association of Clinical Anatomists. 2008. AACA Best Practices for Donor Programs. URL: http://69.167.145.151/images/downloads/aaca_best_practices_for_donor_programs.pdf [accessed 18 April 2015].

Anyanwu GE, Udemezue OO, Obikili EN. 2011. Dark age of sourcing cadavers in developing countries: A Nigerian survey. *Clin Anat* 24:831–836.

Boeckers A, Brinkmann A, Jerg-Bretzke L, Lamp C, Traue HC, Boeckers TM. 2010. How can we deal with mental distress in the dissection room? — An evaluation of the need for psychological support. *Ann Anat* 192:366–372.

Bohl M, Bosch P, Hildebrandt S. 2011. Medical students' perceptions of the body donor as a "first patient" or "teacher": A pilot study. *Anat Sci Educ* 4:208–213.

Bolt S. 2012. Dead bodies matter: Gift giving and the unveiling of body donor monuments in the Netherlands. *Med Anthropol Q* 26:613–634.

Champney TH. 2011. A proposal for a policy on the ethical care and use of cadavers and their tissues. *Anat Sci Educ* 4:49–52.

Chan LK, Pawlina W. (Editors) 2015. *Teaching Anatomy: A Practical Guide.* New York: Springer.

Dickinson GE, Lancaster CJ, Winfield IC, Reece EF, Colthorpe CA. 1997. Detached concern and death anxiety of first-year medical students: Before and after the gross anatomy course. *Clin Anat* 10:201–207.

Drake RL. 2014. A retrospective and prospective look at medical education in the United States: Trends shaping anatomical sciences education. *J Anat* 224:256–260.

Drake RL, McBride JM, Lachman N, Pawlina W. 2014. An update on the status of anatomical sciences education in United States medical schools. *Anat Sci Educ* 7:321–325.

Drake RL, McBride JM, Pawlina W. 2009. Medical education in the anatomical sciences: The winds of change continue to blow. *Anat Sci Educ* 2:253–259.

Dyer GSM Thorndike MEL. 2000. *Quidne mortui vivos docent?* The evolving purpose of human dissection in medical education. *Acad Med* 75:969–979.

Hildebrandt S. 2010. Lessons to be learned from the history of anatomical teaching in the United States: The example of the University of Michigan. *Anat Sci Educ* 3:202–212.

Jones DG, Whitaker MI. 2009. *Speaking for the Dead: The Human Body in Biology and Medicine.* 2nd Edition. Burlington: Ashgate Publishing.

Jones TW, Lachman N, Pawlina W. 2014. Honoring our donors: A survey of memorial ceremonies in United States anatomy programs. *Anat Sci Educ* 7:219–223.

Keniston K. 1967. The medical student. *Yale J Biol Med* 39:346–358.

Kerby J, Shukur ZN, Shalhoub J. 2011. The relationships between learning outcomes and methods of teaching anatomy as perceived by medical students. *Clin Anat* 24:489–497.

Kooloos JGM, Bolt S, van der Straaten J, Ruiter DJ. 2010. An altar in honor of the anatomical gift. *Anat Sci Educ* 3:323–325.

Lachman N, Pawlina W. 2006. Integrating professionalism in early medical education: The theory and application of reflective practice in the anatomy curriculum. *Clin Anat.* 19:456–460.

Lin SC, Hsu J, Fan VY. 2009. "Silent virtuous teachers": Anatomical dissection in Taiwan. *BMJ* 339:b5001.

Louw G, Eizenberg N, Carmichael SW. 2009. The place of anatomy in medical education: AMEE guide no 41. *Med Teach* 31:383–387.

Lucas J. 2013. The dark world of the resurrection men. *MDU J* 29(1):8–18.

MacDonald H. 2005. *Human Remains: Episodes in Human Dissection.* Melbourne: Melbourne University Publishing.

McHanwell S, Brenner E, Chirulescu ARM, Drukker J, van Mameren H, Mazzotti G, Pais D, Paulsen F, Plaisant O, Caillaud MM, Laforêt E, Riederer BM, Sañudo JR, Bueno-López JL, Doñate-Oliver F, Sprumont P, Teofilovski-Parapid G, Moxham BJ. 2008. The legal and ethical framework governing Body Donation in Europe — A review of current practice and recommendations for good practice. *Eur J Anat* 12:1–24.

Martyn H, Barrett A, Broughton J, Trotman P, Nicholson HD. 2013. Exploring a medical rite of passage: A clearing of the way ceremony for the dissection room. *Focus Health Prof Educ* 15:43–53.

Netterstrøm I, Kayser L. 2008. Learning to be a doctor while learning anatomy! *Anat Sci Educ* 1:154–158.

Oxley da Rocha AO, Tormes DA, Lehmann N, Schwab RS, Canto RT. 2013. The body donation program at the Federal University of Health Sciences of Porto Alegre: A successful experience in Brazil. *Anat Sci Educ* 6:199–204.

Pabst R. 2009. Anatomy curriculum for medical students. What can be learned for future curricula from evaluations and questionnaires completed by students, anatomists and clinicians in different countries? *Ann Anat* 191:541–546.

Park JT, Jang Y, Park MS, Pae C, Park J, Hu KS, Park JS, Han SH, Koh KS, Kim HJ. 2011. The trend of body donation for education based on Korean social and religious culture. *Anat Sci Educ* 4:33–38.

Participants, delegates and supporters of the International Conference on 'Cultures of Anatomical Collections' 2012. The Leiden Declaration on Human Anatomy/

Anatomical Collections. URL: http://media.leidenuniv.nl/legacy/leiden-declaration. pdf [accessed 18 April 2015].

Pather N. 2015. The pros and cons of dissection and prosection. In: Chan LK, Pawlina W (Editors). *Teaching Anatomy: A Practical Guide*. New York: Springer, pp. 213–221.

Pawlina W, Hammer RR, Strauss JD, Heath SG, Zhao KD, Sahota S, Regnier TD, Freshwater DR, Feeley MA. 2011. The hand that gives the rose. *Mayo Clin Proc* 86:139–144.

Persaud TVN, Loukas M, Tubbs RS. 2014. *A History of Human Anatomy*. 2nd Edition. Springfield: Charles C Thomas.

Reidenberg JS, Laitman JT. 2002. The new face of gross anatomy. *Anat Rec* (New Anat) 269:81–88.

Richardson R. 2001. *Death, Dissection and the Destitute*. Chicago: University of Chicago Press.

Richardson R. 2008. *The Making of Mr. Gray's Anatomy*. Oxford: Oxford University Press.

Slotnick HB, Hilton SR. 2006. Proto-professionalism and the dissecting laboratory. *Clin Anat* 19:429–436.

Sugand K, Abrahams P, Khurana A. 2010. The anatomy of anatomy: A review for its modernization. *Anat Sci Educ* 3:83–93.

Rosner L. 2011. *The Anatomy of Murders: Being the True and Spectacular History of Edinburgh's Notorious Burke and Hare and of the Man of Science Who Abetted Them in the Commission of Their Most Heinous Crimes*. Philadelphia: University of Pennsylvania Press.

Subasinghe K, Jones DG. 2015. Human body donation programs in Sri Lanka: Buddhist perspectives. *Anat Sci Educ* 8:484–489.

Štrkalj G. 2014. The emergence of humanistic anatomy. *Med Teach* 36:912–913.

Talarico EF. 2013. Change in paradigm: Giving back identity to donors in the anatomy laboratory. *Clin Anat* 26:161–172.

Tschernig T, Pabst R. 2001. Services of thanksgiving at the end of gross anatomy courses: A unique task for anatomists? *Anat Rec* (*New Anat*) 265:204–205.

Winkelmann A, Güldner FH. 2004. Cadavers as teachers: The dissection room experience in Thailand. *BMJ* 329:1455–1457.

Zhang L, Xiao M, Gu M, Zhang Y, Jin J, Ding J. 2014. An overview of the roles and responsibilities of Chinese medical colleges in body donation programs. *Anat Sci Educ* 7: 312–320.

1

BIOETHICAL ASPECTS OF COMMEMORATIONS AND MEMORIALS

D. Gareth Jones

Department of Anatomy, University of Otago, Dunedin, New Zealand
dgareth.jones@otago.ac.nz

ABSTRACT

In considering the ethical dimensions of commemorations, the altruism of the donors emerges as predominant, coupled with which is their informed consent. However, these values will only be expressed if there is trust on the part of the community that anatomists will act with integrity. This is because all involved in the donation process are part of a network of relationships — a community of learning about death and the dead body. Within this community, there needs to be recognition of the crucial input and equal value of each participant — from donors and families to staff, students and researchers. It is values such as these that give to thanksgiving ceremonies their rationale, since they are integral to communities of donation. Since anatomists are part of this community, they have obligations towards the communities from which the donors come as well as responsibility for the bodies they study. It is this common humanity that can be celebrated publicly as anatomists thank their donors and families for their precious gift, acknowledging their dependence upon the goodwill and understanding that lies behind it. This is also a time for families to enter into the drama of death and donation in whatever manner is culturally appropriate for them and their community.

INTRODUCTION

The upsurge of interest in holding ceremonies to acknowledge and thank the families of those who have donated their bodies to anatomical and medical education is to be welcomed. At a minimum it represents a trajectory in the attitudes of anatomists towards recognition that they are an integral part of their communities, and are dependent upon the goodwill of others. It suggests that anatomists recognize that they are more than mere scientists and educators. They are human beings dealing with the remains of fellow human beings, and while their respective roles in the quasi-theatrical performance that constitutes dissection and the study of human body parts are different, they complement one another. Implicitly, these ceremonies recognize that the bodies available for study have been donated for specific purposes, thereby sending out a clear message that they are not the unclaimed bodies of societies' outcasts. This is true regardless of the nature of commemorations from one society and one culture to another, whether they have specific religious overtones, or are decidedly pluralist in nature. It is their existence that is significant, rather than their form.

However, despite these widespread developments, little attention has been given to ethical questions. Even if it is claimed that they are to acknowledge and thank donor and family, this is no more than a fundamental aspiration, important though it is. For instance, whose interests are principally at stake — the deceased, family members, students, or anatomists? Should the form of the ceremonies reflect the views of anatomists and students, or should the wishes of the deceased and/or their families be sought? Should family members be encouraged to play an active role as in some religious cultures? Might answers to these questions determine whether ceremonies are held before as well as on the completion of dissection?

DEFINING TERMS

Various terms are used to convey what is being covered by a ceremony before or after use of the body for dissection. The most commonly encountered terms are commemoration, thanksgiving, ceremony, service, and memorial ("memorial ceremonies," Zhang *et al.*, 2008; Jones *et al.*, 2014; "Convocation of Thanks," Pawlina *et al.*, 2011; "Thanksgiving

Service," University of Queensland, 2015; "cremation/burial ceremony," Lin *et al.*, 2009). While these often appear to be used interchangeably, they all convey the notion of remembrance, and of paying tribute to those who in their death have donated their bodies to a worthy cause, that of medical teaching and research. There are also examples of more permanent memorials, such as plaques (Arraez-Aybar *et al.*, 2014), a "forest" (Zhang *et al.*, 2008), and a Book of Remembrance (University of Queensland, 2015).

Each of the terms is pregnant with ethical meaning. People are remembered for what they have given, and are being thanked for this gift of inestimable value. Their altruism is lauded and celebrated, and their families are thanked for the support they have provided in enabling this gift to be given. In acting in this way they have demonstrated their trust in the anatomical staff and students that they have treated the body with respect and have dealt with it in a manner worthy of the donor's memory. These events also direct our attention to the fact that these individuals made an informed decision to act in this way, and that their bodies did not end up in the dissecting room on account of a legal requirement that overrode their own wishes.

In some cultures and for some individuals the word "service" is appropriate since it encompasses the religious significance of the ceremony. In some instances, there is awareness of the afterlife of the deceased, and of the blessings that donation has bestowed upon the individual (Subasinghe and Jones, 2015). A ceremony prior to use in the anatomy department may be a dedication ceremony, with prayers and chants (Winkelmann and Guldner, 2004). Acceptance of the place of such ceremonies is an ethical requirement in those contexts where they are central to the life of the families and communities.

The variety of ceremonies practiced points to the need for anatomists to be alert to community standards and expectations for the form that ceremonies take, and even in the descriptions by which they are known. One would expect the ceremonies that exist to vary markedly, and to have input from families as well as students and staff. Only in this way will ethical requirements be met, since these would be expected to mirror the views and aspirations of the donors and their families. It is interesting that the notions of "silent mentor" and "silent teacher" have arisen in Eastern contexts (Lin *et al.*, 2009; Atmadja and Untoro, 2012), whereas viewing the bodies of the dead as "first patients" has been more common in Western contexts (Ferguson *et al.*, 2006; Jones *et al.*, 2014).

These comments are all based on the assumption that the bodies have been bequeathed for the primary purpose of medical teaching and research. This raises the question of whether ceremonials can or should be held when the bodies (or even some of the bodies) are unclaimed. Under such circumstances, where does "thanksgiving" fit in? What or who are they "commemorating"? There has been no altruism or donation, and there has been no consent on the part of the individual before death or of family members after death. The ethical drivers in this instance are muted and conflicted, since while the students may still be able to benefit from the availability of the body, it has been reduced to the status of an object rather than a human person to be mourned. "Thanksgiving" in this setting is no more than giving thanks for a dead body, and hence has been robbed of much of its ethical richness by stripping it of its humanity. Under these circumstances, one has to question the legitimacy of holding any form of commemoration or thanksgiving ceremony.

FORMING A COMMUNITY OF LEARNING ABOUT DEATH AND THE DEAD BODY

The basic premise from which I am working is the centrality of the "community of learning about death and the dead body." This immediately places donation and dissection within a context, namely, that of the community of both the living and the dead. The living are represented by the families of the deceased and in some cultures by the religious community and the beliefs that hold the community together. The remains of the one who has died are a reminder of the loss of an integral member of that community, and of the grief being experienced by those who have been left behind. Ethically, this network of relationships is a crucial part of the fabric of the community, and it is this that provides signposts for determining the character of the commemorations that are mounted. Consequently, any commemorations that highlighted the benefits to students but ignored the gift element on the part of the donor and family would fall short ethically.

The words of students to family and friends at the University of Otago commemoration very appropriately express the centrality of relationships (Anatomy Thanksgiving Service, 2015).

> *Thank you first and foremost for your presence here, to allow us to share this moment of gratitude, celebration and experiences with you. I am personally changed by the privilege your loved ones have given my classmates and [me]. In them I saw humility, true service and belief in the future. We thank you and your loved ones for the gift you all have generously given.*
>
>   *It has been an absolute privilege to be able to study anatomy in such a vivid way. We know that we have a responsibility to your loved ones, to make the most of this gift that only some people are brave and compassionate enough to give. These people have given up so much, to give people they don't even know a chance to learn.*

The community of learning about death and the dead body has various facets. These are the bequest itself and the nature of the bequest; the reception of the dead body into the department and the place of the family in this; the dissecting process and the learning of anatomy during this; the return of the remains to the family (ashes); the display of gratitude for the gift of the body (thanksgiving ceremony). While only the last of these activities is the subject of this chapter, each one of them is intimately linked to the others. And each one has ethical dimensions. From this it follows that unethical procedures in one area have repercussions for the other areas. For instance, the use of an unclaimed body and the lack of input from family members cannot be isolated from the dynamics of the dissecting process or what is done with the body following dissection. From this, it follows that one would have to question the meaning of a commemoration event were one to be held.

Implicit within the notion of community is the concept of gift (Campbell, 2009). The gift of the body for studying the structure of the human body and also for encountering death emerges from a willingness on the part of the donor to contribute to the welfare of the community, based on trust in the integrity of those responsible for the body at, and after, death. It is these aspects of the donation and dissection process that are celebrated following use of the body. While they acknowledge the altruism and self-giving of the donor, they also recognize the cohesion of the community within which these processes are being played out. Remove the latter and commemorations lose much of their meaning and rationale.

Therefore, bequest is central. Even though involvement of the family at all stages may be neither desired nor feasible, it would appear to represent the ideal situation. How far this is taken will depend upon the cultural setting, and in some instances its religious symbolism. Meeting with family members prior to the start of the course (Vannatta and Crow, 2007; Crow *et al.*, 2012) is one option through to the presence of family members in the dissecting room (Winkelmann and Guldner, 2004; Atmadja and Untoro, 2012). Regardless of the form taken, the ethical driver is the establishment of a relationship with the family such that members of the family, if they wish, do not feel excluded from knowing that their loved one is being looked after in a "decent" fashion (Human Tissue Act, 2008).

Against this background, it becomes evident that a post-study ceremony is an integral part of the whole and should not be looked at as an add-on, let alone as an optional add-on. In the same way, it can be asked whether the less common pre-study ceremonies are also integral to the whole. If community is as crucial as I have suggested, recognition of its stake in body bequests should start at death. If it is this that leads to the ethical supremacy of bequests, it can be argued that an initial thanksgiving ceremony should be held prior to the use of the body in the dissecting room.

One possible illustration of this is the Maori-based *whakawatea* or "clearing of the way" ceremony at the University of Otago (Martyn *et al.*, 2013), which is held in the dissecting room prior to the first class of the academic year. This ceremony was introduced as a means of enabling Maori students to work with deceased people in the dissecting room. However, in the intervening years many of those who attend are not Maori, and it is now recognized that the ceremony serves a range of benefits. For some students it provides cultural safety, while more generally it encourages respect for the bodies of the donors, and simultaneously gives them an opportunity to consider their attitudes towards death.

This particular ceremony was introduced within a specific cultural context, and hence does not provide a model for use in other contexts. Nevertheless, it is a reminder that cultural and religious elements can all-too-readily be ignored within pluralist societies, and should not be eliminated on the grounds that they do not apply to all students or reflect the belief systems of all donors and their families. Ethical considerations that seek to take account of equality and of divergence in worldviews direct our attention towards the plethora of issues raised by ceremonies, both pre- and post-dissection.

Descriptions of ceremonies are routinely confined to what takes place in them, and while these may be enlightening they pay inadequate attention to the rationale for them. Prior to dissecting and the dissecting room experience, some schools have a lecture and/or tutorials on the ethical underpinnings for respecting the dead body, the bequest process, the dead body as a teaching tool, and contemporary ethical quandaries (Jones and Whitaker, 2009). In some countries, issues raised by for-profit as opposed to not-for-profit companies using human remains will also be relevant (Champney, 2015).

Although coverage of ethical queries may not be thought of as core material for ceremonies, I would argue that in the absence of such coverage ceremonies may degenerate into thoughtless rituals. Since they are held as a means of showing respect and giving thanks for the gift of the bodies, they are replete with ethical meaning. Consequently, in my estimation they are integral to the concept of the community of learning about death and the dead body.

RELATIONSHIP BETWEEN DONATION AND THANKFULNESS

Inherent within the bequest ethos is gratitude for the gift that has been given. From this, it follows that arguments in favor of bequeathing bodies for dissection cannot be separated from arguments in favor of thanksgiving ceremonies. This has not been widely recognized, even by those who have contended vigorously for the use of bequests over against the use of the bodies of the unclaimed, including this writer (Jones and Whitaker, 2012). This reflects a lapse in the ethical literature.

The flip side is that if donated bodies are accepted without expressions of gratitude, the gift itself is demeaned. Such expressions of course need not be simply, or even mainly, via commemorations, but these are the public face of gratitude. Letters of thanks to the family are important and are the bedrock of gratitude, and yet they are private. Donations to a medical school are public and some public expression of gratitude is fitting. After all, body bequests and the use of bodies for teaching and research are legitimized by society, and so concerted attempts should be undertaken to ensure that every aspect of the processes involved are transparent, and in this sense are publicly accountable. The role of public ceremonies is to follow through on this public accountability in a department's dealings with its donors and their families. This is over and above the mechanics of its dealings with the Inspector of Anatomy (in British-based legislatures), where records are kept and checked on the inflow of bodies to the department and the subsequent outflow of bodily remains following dissection or research (Human Tissue Act, 2008).

The driving motive so often for making a bequest of one's body is gratitude to the medical profession, and a desire to assist with medical education and medical understanding. In light of these, it is important to build on them after the event and acknowledge the ways in which the donor may have contributed in some of these ways. This is the conceptual thread linking the two. This does not necessitate a blow-by-blow account of what has been carried out, since the use of a body for teaching may not lead to any noteworthy findings beyond the passing on of knowledge obtained in a particular manner. If the body or a part of the body or organ has been prepared in a special way for further study, this may be noteworthy and could be relayed to family members if they showed an interest in such details.

All such communications have to be handled most judiciously, and the amount passed on will vary from one case to the next. The ethical imperative is not to determine how much information should be given out in any one case, but the importance of seeking to be as transparent as feasible. Only in this way does the department publicly signify its commitment to gratitude and thankfulness. While this is the major driving force, public relations may be a supplementary one in that when others see that the department is acting with integrity and trust they too may be enticed to donate their own bodies.

Much of the thrust of the ceremonies in the West is driven by the students — the recipients of the gift. However, the staff and school are just as much recipients on the assumption that they consider that the students will benefit substantially from the gift. It is apposite, therefore, that staff also have a major part to play both at the time the body is received and prior to burial/cremation. But what about the families on behalf of themselves and also the donor?

The cultural and religious significance of the ceremonies for some donors and their families has already been alluded to. Regardless of the details, the question that arises is whether the donor should be made aware at the time of donation that a ceremony will be held and the general thrust of it, and whether they might wish to have any input into it from their own perspective. Ceremonies in other words represent a variety of interests — donors and their families, students, staff, and possibly community representatives.

The thankfulness aspect also raises the possibility of having permanent memorials on medical school grounds. These take a range of forms (Zhang *et al.*, 2008; Arraez-Aybar *et al.*, 2014; University of Queensland, 2015), but all serve to highlight the gift that has been given to medical education and through this to society. It does not follow ethically that such memorials are obligatory, but it is a further expression of a desire to make explicit thankfulness for the gift that has been voluntarily given. It is also a means of alerting a wider public to the ongoing nature of this gift — from the past through to the future.

SEARCHING FOR ETHICAL VALUES

The previous discussion has been based on a number of foundational values, chief among which are the altruism of the donors, coupled with which is their informed consent. These values are only expressed if there is integrity on the part of those responsible for the bodies, leading to trust on the part of the community. All involved in the donation process are part of a network of relationships, what I have described as a community of learning about death and the dead body. The driving force of such a community emerges from its willingness to recognize the crucial input and equal value of each participant — from donors and families to staff, students and researchers. Each has his/her own part to play in this mutually dependent network of support and learning. It is values such as these that give to the ceremonies their *raison d'etre*; they are integral to communities of donation. Under no circumstances should they be viewed as mere optional add-ons.

Anatomists, therefore, need to recognize that they are part of the human community, with obligations towards those whose bodies they study and towards the communities from which these people come. It is this recognition that binds all together and that allows anatomists to indulge in activities that would otherwise be considered outlandish. It is this common humanity that can be celebrated publicly as anatomists thank their donors and families for their precious gift, acknowledging their dependence upon the goodwill and understanding that lies beneath it. This is also a time for families to enter into the drama of death and donation in whatever manner is culturally appropriate for them and their community.

Commemorations are a fitting antidote to unethical practices that have plagued anatomy in the past (Hildebrandt, 2009), and pathology in more recent times (Royal Liverpool Children's Inquiry, 2001). On so many occasions a lack of gratitude has plagued the discipline, and has been evidence of a sad ignorance of basic ethical values. Thankfully, such times appear to be in the past, and it behoves contemporary anatomists to ensure that the lessons learned over many years are never forgotten. Commemorations serve as salutary reminders of fundamental and ongoing ethical lessons.

REFERENCES

Anatomy Thanksgiving Service. 2015. *Book of Remembrance: Anatomy Thanksgiving Service, Dunedin* 2015. Dunedin: Department of Anatomy.

Arraez-Aybar L-A, Bueno-Lopez JL, Moxham BJ. 2014. Anatomists' views on human body dissection and donation: An international survey. *Ann Anat* 196:376–386.

Atmadja DS, Untoro E. 2012. The usage of the voluntary cadaver in education of medicine through silent mentor program. *Indonesian J Legal Forensic Sciences* 2(2):34–36.

Campbell AV. 2009. *The Body in Bioethics*. Oxford: Routledge-Cavendish.

Champney T. 2015. The business of bodies: Ethical perspectives on for-profit body donation companies. *Clin Anat* 29:25–29.

Crow SM, O'Donoghue D, Vannatta JB, Thompson BM. 2012. Meeting the family: Promoting humanism in gross anatomy. *Teach Learn Med* 24:49–54.

Ferguson KJ, Iverson W, Pizzimenti M. 2006. Constructing stories of past lives: Cadaver as first patient: "Clinical summary of dissection": writing assignment for medical students. *Perm J* 12:89–92.

Hildebrandt S. 2009. Anatomy in the Third Reich: An outline. Part 2. Bodies for anatomy and related medical disciplines. *Clin Anat* 22:894–905.

Human Tissue Act. 2008. Wellington: New Zealand Government.

Jones DG. Whitaker MI. 2009. *Speaking for the Dead: The Human Body in Biology and Medicine*, 2nd Edition. Farnham, Surrey, UK: Ashgate Publishing Ltd.

Jones DG, Whitaker MI. 2012. Anatomy's use of unclaimed bodies: Reasons against continued dependence on an ethically dubious practice. *Clin Anat* 25:246–254.

Jones TW, Lachman N, Pawlina W. 2014. Honoring our donors: A survey of memorial ceremonies in United States anatomy programs. *Anat Sci Ed* 7:219–223.

Lin SC, Hsu J, Fan VY. 2009. Silent virtuous Taiwanese teachers. *BMJ* 339:1438–1439.

Martyn H, Barrett A, Broughton J, Trotman P, Nicholson HD. 2013. Exploring a medical rite of passage: A clearing of the way ceremony for the dissection room. *Focus Health Prof Educ* 15:43–53.

Pawlina W, Hammer RR, Strauss JD, Heath SG, Zhao KD, Sahota S, Regner TD, Freshwater, DR, Feeley MA. 2011. The hand that gives the rose. *Mayo Clin Proc* 86(2):139–144.

Royal Liverpool Children's Inquiry. 2001. *The Royal Liverpool's Inquiry Report*. London: House of Commons.

Subasinghe SK, Jones DG. 2015. Human body donation programs in Sri Lanka: Buddhist perspectives. *Anat Sci Educ* 8:484–489.

University of Queensland, 2015. UQ body donor Thanksgiving Ceremony. https://www.uq.edu.au/sbms/thanksgiving-ceremony (accessed 2 September 2015)

Vannatta JB, Crow SM. 2007. Enhancing humanism through gross anatomy: A pre-course intervention. *Med Educ* 41:1108.

Winkelmann A, Guldner FH. 2004. Cadavers as teachers: The dissecting room experience in Thailand. *BMJ* 329:1455–1457.

Zhang L, Wang Y, Xiao M, Han Q, Ding J. 2008. An ethical solution to the challenges in teaching anatomy with dissection in the Chinese culture. *Anat Sci Educ* 1:56–59.

2

HUMAN FLOURISHING: IMPLICATIONS FOR MEDICINE, EDUCATION AND COMMEMORATION

John McClean

Christ College, 1 Clarence St, Burwood, 2134 NSW, Australia
jmcclean@christcollege.edu.au

ABSTRACT

Like other human focused disciplines, medicine benefits from a robust account of human flourishing. Aristotle's eudaemonistic view, which seeks to describe the *telos* of human life objectively, is a better basis for medical care than a hedonic approach. The chapter reviews several reasons why a eudaemonistic view is preferable, some general and others specific to medical care. It notes Nussbaum's "capability approach" as a more rounded account than Aristotle. The chapter then explores several important elements of an account of human flourishing: religious and spirit engagement, taking moral responsibility, physical health, social relationships, verbal communication, and sharing stories and telling history. These are all relevant to medical care. The conclusion relates the account of flourishing to practices of commemorating those who have donated their bodies for anatomy study.

INTRODUCTION

There is a growing appreciation that reflection on how to best train health professionals requires an understanding of what it means to be human and how humans flourish. The establishment of the University of Durham's Centre for Medical Humanities and University College of London King's College Centre for Humanities and Health, both focusing on health, well-being and human flourishing are exemplars of this trend. Similar concerns can be found in the development of positive psychology (Seligman, 2011) as well as discussions on human rights (Kleinig and Evans, 2013), legal theory (McBride, 2013) and education (Wright and Pascoe, 2014). In each of these domains there are calls for sustained reflection on human flourishing as a basis for determining best practices. Like other human focused disciplines, medicine benefits from a robust account of human flourishing. I will outline some of the history of reflection on human flourishing and the value of the discussion for medical care. I will then sketch one account of human flourishing and note implications for medical care and education. The conclusion draws these threads together by suggesting why body donation and its proper commemoration serves human flourishing and contributes to ethical development in medical training.

THE SEARCH FOR HUMAN FLOURISHING

The question of what is a flourishing human life is an ancient one, going back at least to Socrates, and was the central question of classical Greek philosophy. Aristotle's answer is that human nature has a proper end (*telos*), and a life which reaches such an end is fulfilled (Aristotle, 2012). The term *eudaemonia* denotes for Aristotle the state of flourishing. It is sometimes translated as 'happiness', but for Aristotle it is far more objective than that term suggests. He is not interested in discovering what makes people feel happy (he dismisses pursuing pleasure or enjoyment as "a life suitable to beasts"). Aristotelian flourishing is primarily objective.

So Aristotle examined human nature to determine its *telos*. He concluded that humans are rational, social animals; and we reach our goal as we exercise our nature virtuously, expressing its purpose skillfully. Our animal natures are the lowest and least important element and are satisfied with physical needs: food, rest, shelter and sex. The key to human functioning is reason, both practical or theoretical. The goal of human living is to understand carefully, think clearly and act effectively (reason includes all of that).

So the intellectual virtues of intelligence and wisdom guide the development of other virtues. As social beings, there is a political aspect to our *telos,* we should live together well.

The key to *eudaemonia*, then, is to the develop rational-social virtues which together constitute the full expression of human nature. Aristotle's virtues are far wider than most modern moral "virtues" and include courage, temperance, generosity, bountifulness, pride, good temper, truthfulness, friendliness, leisure and wittiness (Reeve, 2014). Such excellence, in Aristotle's account, always requires thought and understanding. Unreflective popular opinion is no guide to human flourishing (Shields, 2015).

The alternative to Aristotle's eudaemonistic approach is that which assumes that the measure of a flourishing life is subjective, we simply seek pleasure — the *hedonic* view (Haybron, 2008; Feldman, 2010; Raibley, 2012; Huta and Waterman, 2014). This assumes that self-awareness of pleasure is the measure of happiness and that individuals know what will provide them with such pleasure: "Nothing can make you better off that goes against your all-things-considered (informed etc.) preferences, desires or judgment" (Haybron, 2008). Such approaches reach back at least to Jeremy Bentham, the father of utilitarianism, who declared that "quantity of pleasure being equal, pushpin is as good as poetry". That is, there are no 'higher' pleasures, a children's game may be as valuable as sophisticated literature, all that counts is what gives most people the most pleasure. This is quite different to the eudaemonistic approach which aims to determine the proper end of human life and then strives to reach that.

My argument is that medical profession, and any other genuinely ethical activity, needs to be grounded in a vision of the good life which is something like Aristotle's. We should seek a more objective account of the good human life than merely what brings individuals pleasure.

WHY SHOULD WE SEEK AN ACCOUNT OF HUMAN FLOURISHING?

Eudaemonism seems to threaten the autonomy claimed in modern society. Modernity offers freedom by removing any normative account of human nature; and eudaemonism may seem to overturn that. Before turning to positive reasons for preferring eudaemonism over hedonic approaches, it is important to note that a eudaemonistic account does not rule out genuine moral autonomy, understood as the opportunity and requirement to take responsibility for our own lives. Indeed, such requirement can be grounded in a view of human nature, and I will suggest just this below.

To an extent a preference for eudaemonism over hedonism will arise from our worldview, including our religious, or non-religious, convictions. There are, however, some considerations which support eudaemonism. I will briefly deal a few general ones before turning to some which are more specifically medical.

If we recognize that there are significant elements that humans have in common, it is a short step from that observation to the suggestion that all good human lives will have a great deal in common. Aristotle's recognition that humans are social and that living well includes living in good relationships with others is an example of a general feature of a flourishing human life which seems to be widely accepted.

Assertions such as the Universal Declaration of Human Rights presume a shared human nature as the basis of rights. The claims that all humans are "born free and equal in dignity and rights" and entitled to "life, liberty and security of person" can hardly be given a purely hedonic grounding. It is not that we grant these rights to one another because it gives us pleasure. Rather, we recognize that people have rights, which place obligations on others, just because they are human. Rights do not extend to all our preferences, there are specific features of life which should be available to all. One basis on which we determine what should be a 'human right', is to ask what is essential or basic to human living and this is eudaemonistic reasoning. The basis of human rights is a topic of considerable debate, and there is no consensus that rights are best grounded eudaemonistically (Nussbaum, 1997; Wolterstorff, 2008). It is at least clear that they cannot be grounded hedonically.

Haybron notes that the general assumption of nature-fulfilment is, in fact, widely appealing and even underlies desire theories which appear to be hedonic. That is, many approaches which argue that people should be free to seek their own desires, do so because they assume that desires are a reflection of what that person is. Haybron (2008) emphasizes:

> *Eudaemonistic ideals can arguably be found among not just the ancients and their followers but Thomists, Marxists, Nietzsche, the existentialists, and humanistic psychologists like Maslow and Rogers, among many others ... it is questionable whether desire theories of human welfare would be so popular if we did not also tend to think that our desires, understood broadly to include values, ideals and the like, are important to who we are. In satisfying our important desires, we find self-fulfillment.*

Beyond these general considerations, there are some reasons why medicine, in particular, benefits from an account of human flourishing.

Evidence-based medicine (EBM) seeks to base practice on the best clinical outcomes. But what actually constitutes good clinical outcomes? It may seem common sense that the goal of medicine is to treat abnormal and painful symptoms and seek to cure disease, yet it is not always clear what should count as a disease. An account of human flourishing provides a base line for assessing disease.

If we move from disease to health, the need for an account of human flourishing is more evident. The World Health Organization offers a holistic definition of health — "a state of complete physical, mental and social well-being and not merely the absence of disease or infirmity" (WHO, 2006). If medical care aims for such health, then it requires some clarity about what constitutes flourishing and well-being.

Even while treating symptoms, the question of what is a good life inevitably presses in. For instance, the goal of providing dignity, especially at the beginning and end of life, further brings to the fore the need for a clear vision of human flourishing. When patients are unable to express their wishes, care givers must act according to their natural rights which align with some objective good.

A further advantage of having an account of human flourishing is that it can guide research programs. In practice, it is easy for EBM to focus on results which are most easily measured, since these are the most readily available evidence. A vision of a flourishing human life can set goals for medical care which are the result of considered reflection and discussion. These may prove difficult to measure and so hard to investigate evidentially, yet a broader view of the good human life will challenge researchers to seek to examine them, rather than settle for assessing more easily measurable outcomes. On the basis of eudaemonism, EBM can focus on more profound issues of human living. It can ask not simply if a medication reduces blood sugar levels or a therapy increases mobility, but how this contributes more broadly to a flourishing life.

Eudaemonism can help to free patient-centered care from consumerism (Berwick, 2009). Medicine is not best driven by consumer demand in the same way as the fashion or the entertainment industries. There is an extraordinary level of technical knowledge required to make good judgements in the light of modern medical capacities. Few patients are in a position to make deeply informed assessment of possible treatments and outcomes and often want the physicians guidance. "Doctor, what would you do in my situation" is a common question. Assessment of risks and rewards is very difficult, and requires not only technical knowledge, but also a sophisticated view of human flourishing.

There are, then, a range of reasons to think that a eudaemonistic account of human life will serve medical care and education.

DEVELOPING AN ACCOUNT OF HUMAN FLOURISHING

Aristotle's account is not the final word. He acknowledges a connection between ethical living and physical health, but his account of the human *telos* has little to do with physical health. Nussbaum's "capability approach" draws on the Aristotelian tradition but gives a more rounded and satisfying account and offers a basis from which we might develop an approach which would help in medicine (Moody-Adams, 1998; Nussbaum, 2011).

Nussbaum enumerates the following capabilities, that is, opportunities to exercise human powers in a "truly human way":

- life — being able to live to the end of a human life of normal length;
- bodily health, including reproductive health;
- bodily integrity — freedom of movement and physical security;

- being able to use the senses, imagination and thought and to do this in a 'truly human' way;
- emotions — being able to have attachments to things and people outside ourselves;
- practical reason — being able to form a conception of the good and to engage in critical reflection about the planning of one's life;
- affiliation — being able to live with and before each other, and being able to have the social bases for dignity;
- being able to live with concern for other species;
- being able to laugh, to play, to enjoy recreational activities;
- being able to control one's environment through political participation, right to ownership and meaningful work (Nussbaum, 2011).

One of the appealing features of the capability approach is the emphasis on providing opportunities rather than assessing functions. We have good reasons to respect a patient's autonomy, and even to seek to enhance the opportunities to exercise this. Thus, the relevant test for good care is whether the person has appropriate opportunities to exercise various capacities.

Every description of human flourishing is grounded in a philosophical or religious tradition. Such an account cannot be derived merely through empirical studies which report features present in populations or valued by a particular social group, but cannot deliver a normative account of a human life.

The remainder of this chapter offers my reflection on the discussion of human flourishing and its application to medical professionalism.[a]

Religious and Spiritual Engagement

In my view, spiritual questions are basic for human flourishing, and pastoral and clinical care should make space for people to ask and answer

---

[a]My reflection has its base in Christian tradition. You, the reader, may not share these convictions. I offer them as an exposition of the approach of one tradition that can serve as a stimulus for reflection in other traditions.

spiritual questions. Good practice in palliative care has recognized this dimension for a long time. The spiritual dimension is present in all situations, and is often amplified for people facing acute or chronic diseases. There is evidence that religious and spiritual factors are associated with better social and physical health outcomes in cancer patients (Jim *et al.*, 2015; Sherman *et al.*, 2015). Religious and spiritual engagement should not be presumed or required, but the risk in the modern Western health system is that this area will be over-looked rather than required.

Moral Responsibility

Humans are moral beings, and the flourishing life involves the opportunity to exercise this. For many people moral direction is closely related to spiritual and religious dimensions of life. "Autonomy", the capacity to self-determine, has been central to bioethics for five decades (Beauchamp and Childress, 2012). However problematic autonomy may be as a standalone principle, its importance is a recognition that patients (and others) should be treated as morally responsible and enabled as far as possible to make decisions for themselves (Jennings, 2007).

Physical Well-Being

Healthy bodies are an aspect of human flourishing. People live fulfilled, productive, meaningful lives with significant disabilities and diseases because there is a range of dimensions for human flourishing. So a deficit in one area does not result in a necessarily impoverished life. Nevertheless, physical pain, disability and debilitating disease can be a serious obstacle to human flourishing. Allowing patients to overcome these obstacles, in various ways, supports their flourishing. This is the aspect of human flourishing which receives the most attention in most contemporary medical care.

Communal and Social Connections

Humans are communal or social. Our existence, identity, prosperity and destiny are shared, we know ourselves and flourish with others, not alone. This understanding of the importance of community contrasts with the individualism which defines most Western culture. The ideology of consumerism, 'social' media replacing face to face connecting, the demise of voluntary organizations and a host of other factors reinforce individualism (Clarke *et al.*, 2009). The result is that social isolation and loneliness are signification problems in some societies (e.g., Franklin and Tranter, 2011; Wood, 2013). There is evidence that correlates social isolation with poor health and one researcher suggests that loneliness should be ranked with obesity and smoking as a "serious risk factor for poor health" (Cacioppo and Patrick, 2008; Holt-Lunstad *et al.*, 2015). The obvious implication of this is that humans need to have social connection in order to flourish.

So supporting patients and students to flourish means allowing them to engage in social life and connect with a community. It is no surprise that there is a body of literature documenting the importance of social connections for patient outcomes, advocating for hospitals to make it easier for patients to remain connected with their families and communities, and prioritizing the doctor–patient interaction (Charalambous, 2014).

Verbal Communication

Humans are made for verbal communication. Words and language are basic to who we are and how we relate to one another. We engage the world, understanding it and shaping it, by our words. In medical practice, this suggests the importance of communication and conversations. From explaining procedures and laying out options to the warm greeting and gentle reassurance, words make a difference to people.

In contemporary psychiatry, medicalized models are preferred to the "talking cure" of psychotherapy. I'm not promoting a particular school of thought in that debate, and I have already stressed the integration of body and mind. Yet, words matter in the way we treat people — whether we are thinking about medical treatment or day-to-day dealings. Communicating carefully and allowing time to ask questions and discuss are important parts of caring for people well. As Aho and Guignon (2011) put it "the client is not a de-contextualized set of symptoms but an embodied and linguistic way of being whose identity is created and constituted through dialogical relations with others." The medical team cares for people who know themselves and make sense of their lives through words.

Telling Stories and Histories

Humans are storied and historied creatures. This comes from our verbal capacity and our embodiediness in a world which flows in time and our capacity to develop and shape culture. We love stories and we tell stories and we understand ourselves through shared stories. So much of our communication with each other involves telling and listening to stories. From case studies to news articles to movies to jokes — we love narratives. We don't just love them, we need stories to make meaning of our lives and experiences.

Taking the history of a patient has been a standard element of medical practice, though its importance has been reduced with the availability of a wide range of other information to the physician and the rise of evidence-based medicine (Chin-Yee and Upshur, 2015). The importance of stories and histories for human flourishing is a good reason to continue to emphasize the value of history-taking. It does more than establish rapport and elicit basic clinical information. It can help the patient flourish and help the clinician understand what flourishing may be like for this patient (Chin-Yee and Upshur, 2015).

CONCLUSION

The six features considered above are not an exhaustive treatment, but highlight some of the key features of human flourishing which have particular significance in medicine. In concluding these reflections, I point out the connections to practices of commemorating those who have donated their bodies for anatomy study.

Donation, itself, should be related to of human flourishing. When a person determines that their body will be used to help students learn anatomy, they project a new chapter in the story of their body. Their body will remain present to other humans and aid the physical flourishing of others. The capacity to project such a future chapter is an aspect of being storied, and the act of projection adds meaning to life for the donor. As a *donum* (a gift), it is a moral act. "Body snatching" was a desecration of a human body because it used the body without regard to the intention of the person whose body it was (Frank, 1976). Accepting and using a donated body honors the donor and their body.

A fundamental trait of being human is the capacity to remember, acknowledge and reflect. When a person has donated their body, it is 'very human' to commemorate them; to remember and acknowledge a specific life. Such acts of remembering enable us to understand our own lives.

Commemorating a donation can deepen the human significance of the gift. Commemoration will recall the story of the donor, offering family and friends a way to remember the person and recognize their gift. Where appropriate, commemoration should reflect the religious orientation of a donor. It also retains the identity of the donor for a wider social group. Graveyards surrounded medieval churches as a reminder of the 'communion of the saints', those who worshipped in the church knew that they stood among previous generations and shared the same hope of the resurrection. In an analogous way, commemoration reminds those who benefit from the bodies that they receive gifts from those who have gone before to be used for the benefit of a new generation. If this can encourage thankfulness and respect from staff and students, then it enriches the human dimension of donation.

Finally, commemoration engages the moral capacity of medical students, helping to humanize their approach to future practice. They are reminded that, even in the anatomy lab, they are involved in a discipline which demands a high regard for people and their bodies. Involving students in commemoration practices helps to extend medical education into areas of compassion, empathy, caring and personal growth. These are increasingly being realized as essential qualities for a good clinician and that should be taught to medical students in their training (GMC, 2010).

Commemoration of donors can help to establish an important link between lived human experience and clinical care. The art of medicine requires a recognition of the moral and spiritual depths of humanity and the limits of biological possibility. Commemoration is an experience in what it means to care for others in a way that goes beyond the purely physical.

To be human means far more than having a body. A rich account of human flourishing, which includes physical health but extends to a far wider range of capacities, is a foundation for medicine to remain a genuinely human vocation.

REFERENCES

Aho K, Guignon C. 2011. Medicalized psychiatry and the talking cure: A hermeneutic intervention. *Hum Stud* 34(3):293–308.

Aristotle. 2012. *Aristotle's Nicomachean Ethics*. Bartlett RC, Collins, SD trans. Chicago: University of Chicago Press.

Berwick DM. 2009. What 'patient-centered' should mean: Confessions of an extremist. *Health Aff* 28(4):w555-w565.

Beauchamp T, Childress J. 2012. *Principles of Biomedical Ethics*. 7th Edition. New York: Oxford University Press.

Cacioppo JT, Patrick W. 2008. *Loneliness: Human Nature and the Need for Social Connection*. New York: W.W. Norton.

Charalambous L. 2014. Intelligent use of open visiting would aid patient recovery. *Nursing Times* 110(22):11.

Chin-Yee BH, Upshur REG. 2015. Historical thinking in clinical medicine: Lessons from R.G. Collingwood's philosophy of history. *J Eval Clin Pract* 21:448–454.

Clarke GJ, Cameron AJB, Jensen MP. 2009. Towards a Christian understanding of the concept of human "Community", with special reference to the praxis of a non-government human services delivery organization. *ERSP* 3(2):22–40.

Feldman F. 2010. *What is This Thing Called Happiness?* Oxford: Oxford University Press.

Frank JB. 1976. Body snatching: A grave medical problem. *Yale J Biol Med* 49:399–410.

Franklin A, Tranter B. 2011. Housing, Loneliness and Health. AHURI Final Report No.164. Melbourne: Australian Housing and Urban Research Institute.

General Medical Council. 2010. Your Health Matters. GMC, 2010. http://www. gmc-uk. org/doctors/information_for_doctors/7033.asp.

Haybron DM. 2008. Happiness, the self and human flourishing. *Utilitas* 20(1):21–49.

Holt-Lunstad J, Smith TB, Layton JB, 2015. Social relationships and mortality risk: A meta-analytic review. *PLoS Med* 7(7):e1000316.

Huta V, Waterman AS. 2014. Eudaimonia and its distinction from hedonia: Developing a classification and terminology for understanding conceptual and operational definitions. *J Happiness Stud* 15:1427–1428.

Jim HSL, Pustejovsky JE, Park CL, Danhauer SC, Sherman AC, Fitchett G, Merluzzi TV, Merluzzi TV, George L, Snyder MA, Salsman JM. 2015. Religion, spirituality, and physical health in cancer patients: A meta-analysis. *Cancer* 30:3760–3768.

Jennings B. 2007. Autonomy. In: Steinbock B (Editor). *The Oxford Handbook of Biotheics*. Oxford: Oxford University Press, p. 72–89.

Kleinig J, Evans NG. 2013. Human flourishing, human dignity, and human rights. *Law Philos* 32:539–564.

McBride N. 2013. Tort law and human flourishing. In: Pitel SGA, Neyers JW, Chamberlain E (Editors). *Tort Law: Challenging Orthodoxy*. Oxford, Portland: Hart Publishing, p. 19–57.

Moody-Adams MM. 1998. The virtues of Nussbaum's essentialism. *Metaphilosophy* 29:263–272.

Nussbaum M. 2011. *Creating Capabilities: The Human Development Approach*. Cambridge: Belknap/Harvard University Press.

Nussbaum M. 1997. Capabilities and Human Rights. Fordham *L Rev* 66:273–300.

Raibley J. 2012. Happiness is not well-being. *J Happiness Stud* 13:1105–1129.

Reeve CDC. 2014. Beginning and ending with Eudaimonia. In: Polansky R (Editor). *The Cambridge Companion to Aristotle's Nicomachean Ethics*. New York: Cambridge University Press, p. 14–33.

Seligman MEP. 2011. *Flourish: The New Positive Psychology and the Search for Well-Being* New York: Free Press.

Sherman AC, Merluzzi TV, Pustejovsky JE, Park CL, George L, Fitchett G, Jim HSL, Munoz AR, Danhauer SC, Snyder MA, Salsman JM. 2015. A meta-analytic review of religious or spiritual involvement and social health among cancer patients. *Cancer* 30:3779–3788.

Shields, C. 2015. Aristotle. In: Zalta EN (Editor). *The Stanford Encyclopedia of Philosophy*, Fall 2015 Edition. http://plato.stanford.edu/archives/fall2015/entries/aristotle/.

Wolterstorff N. 2008. *Justice: Rights and Wrongs*. Princeton and Cambridge: Princeton University Press.

Wood S. 2013. All the lonely people. Sydney Morning Herald, September 5.

Wright PR, Pascoe R. 2014. Eudaimonia and creativity: The art of human flourishing. *Cambridge J Educ* 45:295–306

World Health Organization. 2006. Constitution of the World Health Organization. www.who.int/governance/eb/who_constitution.en.pdf.

3

THE LACK OF MEMORIAL SERVICES IN FOR-PROFIT AND NON-PROFIT WILLED BODY COMPANIES IN THE UNITED STATES

Thomas H. Champney

Institute for Bioethics and Health Policy & Department of Cell Biology, Miller School of Medicine, University of Miami, USA
tchampney@med.miami.edu

ABSTRACT

In the United States, the majority of university-based or state-based willed body programs provide a memorial service for the users of the bodies and / or for the families of those individuals who donated their bodies. On the other hand, for-profit and non-profit willed body companies rarely provide any type of service. The differences between the various types of willed body organizations in the United States are examined and a discussion comparing these organizations is highlighted. These differences are especially apparent when examining issues such as the availability of memorial services.

INTRODUCTION

In many countries, individuals have, for decades, willingly and altruistically donated their bodies for use in medical education and research (Cornwall *et al.*, 2012). In the United States, the oldest willed body programs are found in non-profit academic institutions or government-sponsored anatomical boards (Champney, 2011; Garment *et al.*, 2007). Over the past two decades, a new business entity has emerged; non-profit and for-profit willed body companies that solicit body donors and then provide these bodies or parts of these bodies to their clients (Anteby and Hyman, 2008). These companies charge their clients for shipping, handling and processing the bodies, but they do not charge them for the actual donated bodies, as this would be against the law. Non-profit companies charge these fees to cover their expenses while for-profit companies charge higher fees enabling a profit to be generated (Champney, 2016). In addition, for-profit plastination companies have been formed that solicit body donors for use in exhibitions as well as to sell plastinated parts to educational institutions (Jones, 2000; Tanassi, 2007).

BACKGROUND OF UNIVERSITY-BASED WILLED BODY PROGRAMS

With the development of medical schools in the United States, a need arose for human bodies to be used for teaching normal anatomy to the medical students. To acquire these bodies, universities established local willed body donor programs that would solicit donors from the community to will their bodies to the medical school for use in teaching students or for use in medical research (Garment *et al.*, 2007). These programs are non-profit, are administered by the university or by a state-based anatomical board and generally use funding from the university or the state to pay for overhead to run the program (Schmitt *et al.*, 2014). The individuals donating their bodies typically arrange the donation many years before their death and they make these arrangements with the knowledge that their donation is helping the local medical school (Anteby and Hyman, 2008). The donors do not receive any payment for their donation, but, by donating, the donors have minimal funeral expenses which in the majority of cases are covered by the willed body program.

There are dozens of university and state-based willed body programs in the United States with the majority of these programs working in their local communities. Virtually all of these programs have a memorial service to honor the donors (Jones *et al.*, 2013). This service includes the students who gained valuable knowledge and experience from the bodies and may also include family members of the donors. In many cases, the administrators of the programs are also those that teach from the willed bodies, therefore the ability (and desire) to have a memorial service is easily accomplished. The willed body programs also return the cremated remains of the donors to their families or distribute them in a dignified and respectful manner. These willed body programs are locally-based and handle all aspects of the donation process from inquiry, through sign up, to collection of the body after death and disposition of the cremated remains.

BACKGROUND OF NON-PROFIT AND FOR-PROFIT WILLED BODY COMPANIES

Over 20 years ago, it was realized that there was a need for more cadavers than the local willed body programs could provide. Many of these needs were from surgical training companies that used human body parts to train surgeons in the use of newly developed surgical equipment, such as arthroscopy or other forms of minimally-invasive surgery. Other needs included refresher certification courses for physicians or undergraduate schools that wanted to teach human anatomy but did not have access to a willed body program. At first, non-profit willed body companies, such as Anatomy Gifts Registry, were established that solicited body donors for medical education and research. A few years later, for-profit companies were established (e.g., Science Care, MedCure, BioGift) to provide the same service as the non-profit, but in different locales. These companies solicit donors throughout the United States from hospices, nursing homes and through advertisements in local media. Many of these donors are near the end of their lives (Anteby and Hyman, 2008) and, after death, the company collects the body, processes the body and ships whole bodies or body parts to training sites for specific use by their clients. After use, the bodies or body parts are returned to the company for cremation and disposition of the remains. Since it is illegal to sell human bodies, these companies generate revenue by charging shipping, handling and processing fees (Champney, 2016).

Since the end users of the human tissues are not the same as the administrators of the willed body companies, there is less likelihood that memorial services are held (Benninger, 2013; Jones, *et al.*, 2013). When four of the most popular and well-known companies were contacted (Science Care, MedCure, BioGift and Anatomy Gifts Registry) and asked whether they provided any type of memorial service, the response was negative. Likewise, the websites of these companies do not mention any type of memorial service. However, another author found that six out of twenty "private procurement organizations" did offer memorial services (Benninger, 2013), although the names of these companies were not provided. It should be noted that some of the end users of the bodies have been known to sponsor memorial services of their own (e.g., the Israeli Rhinology Society), but this is not a common occurrence.

These companies do have other ways to honor the donors. Science Care donates a tree to the Arbor Day Foundation as part of their "Memory in Nature" program and the majority of the non-profit and for-profit companies have online testimonials from the families of donors or the donors themselves. In addition, many of these companies provide general information to the families about the use of their loved one and they will return the cremated ashes, if requested.

It should be pointed out that many of the for-profit body companies have surgical training centers affiliated with their business, so that a user can organize a training session with lecture halls, laboratories and bodies provided at the same site. This provides convenience to the user and would make it easier to have recognition of the willed body donors at these sites.

A different type of for-profit willed body company is the plastination company (e.g., Body Worlds) that solicits donors to have their bodies impregnated with plastic resin, dissected and put on display in traveling exhibitions (Jones, 2016). These companies generate profit by charging admission to the exhibits. The plastinated specimens remain preserved for many years and can be displayed repeatedly. Other plastination companies use donor material to develop plastinated body parts which are then sold to schools and universities as teaching aids. The majority of these companies also provide testimonials on their websites as well as acknowledging the donors' altruism and generosity. They do not, however, offer any type of memorial service for the donors' families.

COMPARISON OF UNIVERSITY-BASED WILLED BODY PROGRAMS WITH FOR-PROFIT WILLED BODY COMPANIES

One reason for the difference in prevalence of memorial services between willed body programs and willed body companies may be due to the nature of these two organizations. The willed body programs are locally-run and generally solicit donors for specific uses in medical training programs. The administrators of the programs are commonly the same individuals who teach the material to the students and who personally understand and appreciate the actual use of the donor material. They are educators and researchers who run willed body programs to fulfill an educational need.

On the other hand, the willed body companies appear to have a more business-like approach. They realize that there is a market in human tissue (Anteby, 2010; Anteby and Hyman, 2008). They solicit donors as products to be supplied to clients. They do understand the emotional and moral nature of their work, but they approach it in a more business-like manner. They are more distant from the end users of the donated bodies and may not recognize or appreciate the need for memorial services. They may also consider that the end users are able to offer a memorial service themselves and that their role is to merely procure and transfer the willed bodies to the users.

These different types of willed body organizations coexist in the United States for a number of reasons. First, there is a need for human material in education, research and medical businesses. While willed body programs can support the majority of educational needs, the non-profit and for-profit companies help fulfill the other needs. In addition, the willed body programs and the willed body companies appear to draw their donors from different donor populations (Anteby and Hyman, 2008), so there may not be direct competition between these two types of organizations, but actually two separate, parallel systems. Second, the regulations and policies for the handling of willed bodies in each of the fifty states in America are not as rigorous as in other countries (Champney, 2016; Charo, 2006; Harrington and Sayre, 2006; Jones, 2000; Zhang, *et al.*, 2013). This allows for the growth and development of a wide range of willed body organizations from university-based willed body programs through for-profit willed body companies. Therefore, it is not surprising that these coexist in the United States and that they could have markedly different approaches to the use of willed bodies and, more specifically, to the availability of memorial services.

CONCLUSIONS

Non-profit and for-profit willed body companies appear to approach the use of donated bodies in a different manner in comparison to the majority of university or government-sponsored willed body programs. The presence of memorial services for those who donate their bodies to these companies is one aspect that highlights the difference between these organizations. The lack of memorial services in for-profit and non-profit companies may serve as an indicator of the different approach taken by these companies compared to the more typical university-based willed body programs.

REFERENCES

Anteby M. 2010. Markets, morals, and practices of trade: Jurisdictional disputes in the U.S. commerce in cadavers. *Adm Sci Q* 55:606–638.

Anteby M, Hyman M. 2008. Entrepreneurial ventures and whole-body donations: A regional perspective from the United States. *Soc Sci Med* 66:963–969.

Benninger B. 2013. Formally acknowledging donor-cadaver-patients in the basic and clinical science research arena. *Clin Anat* 26:810–813.

Champney TH. 2011. A proposal for a policy on the ethical care and use of cadavers and their tissues. *Anat Sci Educ* 4:49–52.

Champney TH. 2016. The business of bodies: Ethical perspectives on for-profit body donation companies. *Clin Anat* 29:25–29.

Charo RA. 2006. Body of research — Ownership and use of human tissue. *N Engl J Med* 355:1517–1519.

Cornwall J, Perry GF, Louw G, Stringer MD. 2012. Who donates their body to science? An international, multicenter, prospective study. *Anat Sci Educ* 5:208–216.

Garment A, Lederer S, Rogers N, Boult L. 2007. Let the dead teach the living: The rise of body bequeathal in 20th-century America. *Acad Med* 82:1000–1005.

Harrington DE, Sayre EA. 2006. Paying for bodies, but not for organs. *Regulation* 29:14–19.

Jones DG. 2000. *Speaking for the Dead: Cadavers in Biology and Medicine.* Aldershot: Ashgate.

Jones DG. 2016. The public display of plastinates as a challenge to the integrity of anatomy. *Clin Anat* 29:46–54.

Jones TW, Lachman N, Pawlina W. 2013. Honoring our donors: A survey of memorial ceremonies in United States anatomy programs. *Anat Sci Educ* 7:219–223.

Schmitt B, Wacker C, Ikemoto L, Meyers FJ, Pomeroy C. 2014. A transparent oversight policy for human anatomical specimen management: The University of California, Davis experience. *Acad Med* 89:410–414.

Tanassi LM, 2007. Responsibility and provenance of human remains. *Amer J Bioeth* 7(4):36–38.

Zhang L, Xiao M, Gu M, Zhang Y, Jin J, Ding J. 2013. An overview of the roles and responsibilities of Chinese medical colleges in body donation programs. *Anat Sci Educ* 7:312–320.

4

REMEMBERING THE VICTIMS OF ABUSIVE PRACTICES IN ANATOMY: THE EXAMPLE OF NAZI GERMANY

Sabine Hildebrandt

Boston Children's Hospital, Harvard Medical School, Boston, USA
sabine.hildebrandt@childrens.harvard.edu

ABSTRACT

Since the late 1980s, investigations into the history of medicine practiced in Germany from 1933 to 1945 have revealed the deep involvement of clinicians and biomedical researchers in the atrocities of the National Socialist (NS) regime. However, only recently has the role of anatomists in the physical destruction of the bodies of perceived enemies of the NS government been studied.

The bodies of victims of NS persecution, that became available within the traditional sources of anatomical body procurement, were used extensively and eagerly by anatomists for teaching and research purposes. The remains — or ashes after cremation — were interred in anatomy cemetery plots, which were usually anonymous. A current focus of research is the identification and restoration of the biographies of these victims, who have remained un-named over many decades. Several different ways of memorialization have been chosen so far, including the publication of names and biographies in scientific journals, books and on websites. Another option could be a memorial service at medical schools for victims of abusive practices in medicine and anatomy.

The National Socialists' attempt to not only physically annihilate these human beings, but also erase their existence from the memory of mankind, shall be revoked by the restoration of the victims' names and biographies.

"Forgetting them would be the victims' final annihilation"
Hans-Joachim Lang, *Die Namen der Nummern*

INTRODUCTION: HISTORICAL AWARENESS

Within the field of ethics in anatomy one of the central topics is the recognition of the anatomical body as the embodied memory of the person once inhabiting it. The anatomical use of this body becomes a relevant part of the person's biography (Winkelmann, 2007; Winkelmann and Schagen, 2009). This conceptualization of the anatomical body is a historically new one, and developed as anatomists learned to ask about the history of the persons whose bodies they are dissecting. Traditionally, scholars of anatomy presented a pervasive "lack of general interest in the source of the bodies" (Jones, 2011). One of the many consequences following this conceptual shift is the realization that a public acknowledgement is owed to the persons whose bodies are used for anatomical dissection. Thus most medical schools who are working with bodies donated within bequest programs now pay tribute to the donors, and honor them with special memorial services and monuments (e.g., USA: Jones *et al.*, 2014; Germany: Pabst and Pabst, 2006; Thailand: Winkelmann and Güldner, 2004; Taiwan: Lin *et al.*, 2009; Sri Lanka: Subasinghe and Jones, 2015). Expressions of gratitude often include the recognition of the donors and their families by name.

But what if the bodies were not obtained through voluntary donation, or even worse: what if the bodies were used for anatomical purposes against the explicit wishes and expectations of the bodies' owners? And what if the names of these persons seem irretrievably lost?

Over the centuries, anatomists have often been involved in the use of bodies of the disenfranchised within a population, those who were discriminated against for so-called "racial", political or economic reasons

(Richardson, 2000; Sappol, 2002; Halperin, 2007; Kenny, 2013). However, a unique situation of abusive practices in anatomy evolved in Germany between 1933 and 1945. During the National Socialist regime (NS, Nazis) anatomists became complicit in the complete annihilation of the perceived enemies of the Nazis, among them persons persecuted for so-called "racial" and political reasons (Hildebrandt, 2014a). Not only did anatomists use these victims' bodies for their scientific and educational purposes, but they refused to acknowledge any wrongdoing for many decades after the war, sometimes even continuing the work with their wartime collections. It is only in the last years that an awareness of the need to take on responsibility for this history has been growing among German anatomists. Historians and anatomists have taken first steps to retrieve the memory of these victims. Whereas it is impossible to undo past wrongdoings, it is possible to pay tribute to the victims and their memory.

WHAT HAPPENED

After the National Socialists came into power in January 1933, one of their first goals was a centralization of the leadership of universities in the hands of the ministry of education. This included anatomical institutes and body procurement, and in the case of bodies of prisoners the responsibility for their delivery was shared with the ministry of justice. Existing legislation concerning body procurement was reaffirmed, and new NS laws led to an exponential increase of executions, thus necessitating regulations that specified the exact distribution of the bodies of the executed to individual universities (Noack and Heyll, 2006; Hildebrandt, 2008). Most of those anatomists who had not been dismissed for so-called "racial" or political reasons pledged their allegiance with the new regime by joining the NS party. Many did so only pro forma, but several were active NS proponents. All of them, independent of their political convictions, used the bodies of victims of the NS regime for their work (Hildebrandt, 2016). They were not only passive recipients of this body supply, but often lobbied for the bodies and competed for the most "desirable" among them, those of executed persons (Noack and Heyll, 2006). In a practice that had been established before 1933, "material" from the executed was preferred by German anatomists, as the tissues could be removed directly after death, were thus fresh, and hailed from usually healthy persons (Hildebrandt, 2013a). "Material" from bodies of the executed was also used to prepare histological specimens for microscopy in the education of medical students, and the students encountered bodies with the physical signs of abuse, decapitation or hanging in the dissection course (Hildebrandt, 2016).

Based on currently available data concerning the body procurement at German anatomical institutes between 1933 and 1945, a first conservative estimate of the total number of bodies used for anatomical purposes arrives at 30,000 to 35,000 bodies for the 31 anatomical departments in Germany and the occupied or annexed territories during the war (Hildebrandt, 2014b). These bodies came from traditional sources of body procurement, among them unclaimed bodies from psychiatric institutions, suicides, hospitals, prisons and executions. The term "unclaimed" refers to bodies of persons whose families did not claim them for burial. These traditional sources changed under the NS regime to include ever increasing numbers of NS victims. Among the dead from the psychiatric hospitals were victims of so-called "euthanasia"; among the suicides were desperate Jewish citizens; among the prisoners were inmates from camps for prisoners of war, forced laborers and concentration camp inmates who died in higher numbers due to the increased violence and harsh conditions of incarceration; and the number of the executed reached previously unknown heights and included women, even some pregnant women. A documented minimum of 3,887 bodies of the executed were delivered to 20 anatomical institutes (Hildebrandt, 2016). The final numbers of the executed will probably lie much higher, and the number of all NS victims among the estimated total is unclear. Schönhagen assumed that two thirds of the total bodies supplied to the anatomy in Tübingen stemmed from NS victims (Schönhagen, 1992).

Among the NS victims whose bodies were delivered to anatomical departments, executed or otherwise put to death, were Jewish citizens, political dissidents of many nationalities, religious dissenters, conscientious objectors, deserters, forced laborers and criminals as defined by NS laws. The anatomists used the bodies of these victims for educational purposes in dissection courses and the teaching of histology, for the preparation of exhibition specimens in anatomical collections, for the creation of anatomical textbooks and atlases, and for hundreds of research studies (see, Hildebrandt, 2016; on textbooks and atlases: Hildebrandt, 2006; Czech, 2015; on research: Hildebrandt 2013a). Even after the war, bodies of NS victims were used at several anatomical institutes for many years longer (Hildebrandt, 2014b; Czech, 2015). The remains of the victims — or ashes after cremation — were interred in anonymous cemetery plots owned by the anatomical institutes. Relatives of victims were not informed, leaving some families searching for information about their loved ones' burial place for decades after the war (e.g., File Gedenkstätte Deutscher Widerstand GDW/P Schubert, Ruth).

IDENTIFICATION OF VICTIMS AND THEIR REMAINS

The scientific community owes a duty of care to the victims, as expressed by a quest to identify and commemorate them as individual persons. All commemoration has to be based on identification, of which Paul Weindling states: "Only by knowing an individual name can the circumstance of the individual's life be reconstructed. Anonymization perpetuates an oppressive stigma" (Weindling, 2013). In view of the total number of bodies delivered to the anatomical departments during the Third Reich, the task of the identification of all or most of the NS victims whose bodies were used for anatomical purposes seems daunting. However, Hans-Joachim Lang has shown that even in dire situations information can be found when the right motivation is there. In painstaking research he was able to identify and reconstruct the biographies of the 86 victims of anatomist August Hirt's human experiment, the so-called "Jewish skeleton-collection", even though initially only their Auschwitz tattoo numbers were known (Mitscherlich and Mielke, 1947; Lang, 2007, 2013a,b). Other identification efforts have concentrated on victims executed by court order, as they are the most likely group to have left a trail of documentation. A list of names of executed victims transported to the Bonn anatomical department was published in 2006, as were names for Berlin and Leipzig and other universities (Forsbach, 2006; Hildebrandt 2013c, 2014b). These studies publicized the names, whereas this was not the case in Jena and Vienna, where names were also identified but not included in the published studies (Jena: Redies *et al.* 2005, 2012; Vienna: Angetter, 1998; overview of currently available data: Hildebrandt 2014c, 2016). So far, about 500 biographies have been reconstructed in more or less detail.

It will be much more difficult to find information on victims from psychiatric institutions and the various camps, especially in the cases of anatomical departments were body registers no longer exist, or are incomplete, or do not contain names. The anatomical institute of Würzburg is an example, where a body register exists, but does not include the names of 80 apparent victims of "euthanasia", whose bodies were delivered in 1941 or 1942 (Blessing *et al.*, 2012). Also, documents often were deliberately falsified in the case of murdered "euthanasia" patients for whom natural causes of death were "invented", a fact that makes them indistinguishable from psychiatric patients who truly died of natural causes. Nevertheless, sometimes it is possible to identify victims through other sources. An example is the detailed work of historian Gunnar Richter, that led to the identification of 24 names of forced laborers who died or were executed at the internment camp of Breitenau near Kassel/Hessen. Their bodies were transported to the anatomical institute at the University of Marburg (Richter, 2009; Hildebrandt, 2016).

An additional duty of identification concerns the possible continued use of specimens derived from bodies of NS victims. Here much work still needs to be done. Whereas most German and Austrian anatomical collections were officially investigated in view of this question, some of these studies were more thorough than others, and most lacked an emphasis on identification of the victims from whose bodies these specimens were taken (Hildebrandt, 2014b). In fact, many anatomical institutes saw these investigations more as an exercise in "cleansing" their collections from potential remnants of a murky past, rather than as a true investigation and acknowledgement of responsibility for this history and a chance to honor the victims (Schönhagen, 1994; Weindling, 2012; Seidelman, 2012). Many collections in Germany and Austria still need to be investigated in a systematic and appropriate fashion that includes outside observers. Also, specimens from the NS era are still surfacing in sometimes completely unexpected locations and are not always appropriately handled (Max-Planck-Gesellschaft 2015; Schmelcher, 2015; Bever, 2015). Another problem that has not been addressed at all are the histological specimens that were given to medical students in the 1930s and 40s. At the University of Berlin, where student-numbers were as high as 1000 per year during the war, sets of 160 slides were handed over to each student to keep as their own (Hoffmann, 1951; Romeis, 1953). Ideally, all specimens potentially hailing from bodies of NS victims should be located, identified, and then buried in an appropriate manner that acknowledges the individual victim.

COMMEMORATION

This leads to the issue of possible ways of memorializing the victims, whose bodies were used for anatomical purposes during the Third Reich. Opinions have generally differed over the last decades as to the most dignified and suitable remembrance. Ultimately they all have to be based on detailed research into the true facts of this history, which alone can provide the foundation for an honest and worthwhile commemoration.

An example of a particularly gracious memorial is the burial for neuroanatomical specimens that had been taken from the bodies of children at the *Spiegelgrund* hospital in Vienna. Here each child is represented individually (Czech, 2002; Weindling, 2013; Kaelber, 2014). Other ways of remembrance include research and publications on individual biographies of perpetrators and victims. Several organizations have also put together memorial statements, listing the names and biographical information. These memorials have varying formats, ranging from review articles, to books like Lang's report on Hirt's victims, and general statements of responsibility by different groups of physicians (review articles: e.g., Hildebrandt and Aumüller, 2012; books: e.g., Lang, 2007; victims executed at Plötzensee: Perk and Desch, 1974; victims murdered at Terezín: Steinhauser, 1987; statements: e.g., for all German physicians: Kolb *et al.*, 2012; for psychiatrists: Schneider, 2011).

A more interactive and comprehensive approach to commemoration is possible through online databanks. Paul Weindling and his colleagues are currently establishing such a data collection for all victims of human experiments and coercive research in National Socialism (see, http://www.history.brookes.ac.uk/research/centres/hms/vhens/). An interactive database for the NS victims in anatomy could contain all personal information of a victim, including name, age, gender, nationality, and profession, as well as facts of imprisonment or hospitalization and other conditions that led to the delivery of their bodies to an anatomical institute. Information on the fate of the victim's body could be noted, including the final resting place. The advantage of a database lies in the ease of adding information as it becomes known, ready availability for analysis of group data, and the possibility of ultimately making it public, similar to the website that Lang has dedicated to Hirt's victims (Lang, 2013a). Another impressive example of an online database has been created by the University of Vienna for the victims and survivors of NS persecution among its employees. It is called a "memorial book" and emphasizes the importance of the victims' remaining "part of the collective memory of today's University" (see, http://gedenkbuch.univie.ac.at/index.php?id=435& no_cache=1&L=2&id435=&no%28underscore%29cache1=&L2=). Similarly, the victims of anatomy could become part of the collective memory in German medicine.

Whereas the creation of a database for victims of anatomy is highly desirable and may be the best way to honor their memory, there are practical problems to consider. Questions concerning privacy protection laws have been raised in connection with the naming of victims of the NS regime, especially with respect to victims of the "euthanasia" crimes, whose families might have felt stigmatization through association with psychiatric diseases. In addition, archival laws of data protection have been interpreted in different manners by archives (Weindling, 2010). However, even in the case of the "euthanasia" victims these problems are not insurmountable any longer, as the example of a recent exhibit organized by historian Astrid Ley for the "euthanasia" victims in Brandenburg showed. After discussions with archives and families, Ley and her colleagues were able to include the full names of victims in the 2012 exhibit (Gedenkstätten Brandenburg/Havel: http://www.stiftung-bg.de/doku/ neues/neues_m1.htm). With respect to the NS victims in anatomy, the archival laws on statutes of limitations for publication of personal data most likely do not apply any longer, as these persons have died more than 60 years ago (see, Bundesarchivgesetz, §5). It should also be noted, that neither the publication of the Forsbach list of names of execution victims in Bonn, nor Lang's naming of Hirt's victims, nor the listing of the women on Berlin anatomist Stieve's list resulted in any negative repercussions from the public (Forsbach, 2006; Lang, 2007, 2013a, b; Hildebrandt, 2013a). One important provision for a databank for purposes of commemoration has to be that, while providing full names and biographical information, the connection of verdicts from court proceedings or medical diagnoses with names of victims has to be reviewed for each case.

One approach to the memorialization of victims of anatomy could be particularly appropriate: the linking of a physical "monument" with an online memorial, somewhat similar to the *Stolpersteine-Projekt* (stumbling blocks project) existing in many German cities (Stolpersteine, http:// www.stolpersteine.eu/en/). In the *Stolpersteine* art project small bronze plaques with basic biographical information of a victim are placed in the pavement in front of the victim's last known place of residence. Each of the plaques is linked to an online-biography. In a similar fashion, anatomical institutes could display physical monuments with the names of NS victims whose bodies were used at this particular institute and a website displaying biographical information on the victim. However, such a project might be difficult to pursue due to potential local resistance against an open association with this history, as well as due to the need for adequate funding. At this moment, some institutes display memorial plaques commemorating the victims in an anonymous fashion. The one at the University of Jena is placed in the entrance hall of the institute and reads "In memory of the victims of National Socialism, whose bodies were delivered to the anatomical institute of the University of Jena between 1933 and 1945" (Redies *et al.*, 2012).

In addition to all of these ways of commemoration, an annual memorial service of remembrance for the victims of NS medicine has been discussed by several authors (Peiffer, 1991; Seidelman, 1996; Cohen and Werner, 2009). A burial of specimens alone might be problematic, as the neurologist Jürgen Peiffer pointed out in a speech given at the first anniversary of the erection of a stone commemorating the victims of NS medical atrocities in a Tübingen cemetery. Peiffer remarked: "There is a dangerous possibility that we may bury our bad consciences together with [...] tissue remains, thereby avoiding the necessity of remembering the past at least once every semester, together with the students" (Peiffer, 1991). Instead, such a ceremony should be held by medical professionals not only in Germany but across the world, to remind the global community of health workers of the legacy of this history. It could facilitate, in Peiffer's words, the willingness of "facing ourselves, our own history, and asking the question whether science, students, and academic teachers are prepared to learn from the past in order to influence the future" (Peiffer, 1991).

CONSEQUENCES

The systematic identification and memorialization of individual NS victims whose bodies were used for anatomical purposes has started only recently. One of the reasons for this late exploration of the victims' biographies lies in the historical reluctance of anatomists to approach the topic of potential wrongdoing in their own profession. At the core of this reluctance was the unwillingness or inability of anatomists active during the Third Reich to admit any moral failure in the use of bodies of victims of the NS regime. As these bodies, including those of the executed, were part of the legal body procurement, they saw no reason to doubt the righteousness of the use of these bodies for teaching and research purposes (Hildebrandt, 2013b). Most of the anatomists continued in their professional positions and attitudes after the war. Their pupils, the next leaders of anatomy, were loyal in the defense of their teachers' actions, and actively opposed any examination of this past until the 1990s (Professor Gerhard Aumüller, personal communication). It is only the latest generation of German anatomists who have started to investigate this history in a systematic fashion and finally organized a first symposium on the history of anatomy in the Third Reich in 2010 (Hildebrandt, 2011). The reasons behind this emerging willingness to confront the past still need further exploration. At least two potential contributing factors can be identified: the general openness of the younger generation of Germans towards a dispassionate examination of the NS past, as well as the change in attitude among anatomists concerning the conceptualization of the anatomical body, that now also includes the history of the person once inhabiting the body. The latter may indicate a more general change in the professional ethics of anatomy during the last century. The examination of the NS history of anatomy is indeed of significance for contemporary anatomy and medicine, as "a legitimate ethical foundation in medicine cannot be established until the profession has demonstrated the insight and capacity to acknowledge evil, to recognize its victims, and to commemorate their suffering" (Seidelman, 1989).

The *Anatomische Gesellschaft* (anatomical society), an organized body of German and international anatomists based in Germany, has honored the "often nameless victims of [NS] justice" who "were robbed of their human dignity" by becoming "material" for anatomical purposes, in publishing a memorial page in a book commemorating the society's 125-year anniversary (Hildebrandt and Aumüller, 2012). This commemoration can only be a first step, as it does not include the names of victims. In a separate statement from 2014, the AG has formally acknowledged in a publication on its website its responsibility for the history of anatomy in the Third Reich, and promised to support any further research on the subject (see, http://anatomische-gesellschaft.de/index.php?id=geschichte-der-anatomie-im-dritten-reich). It would be appropriate, if the website of the AG could also become the home for a database of the victims and their biographies. It would enable its users to learn about the individual victims behind the vast number of persons affected by NS policies, and enable them to "understand what [they] should know about the Nazis", as "our imagination cannot count" (Remarque, in Van Gelder, 1946). At the same time, a database could support a more general commemoration by anatomists and give them, in Professor Seidelman's words, a chance for "recollection", "remembrance" and "a tribute to lives past" (Seidelman, 1999). This is essential as "remembrance will not bring anybody back to life, but it keeps the murdered victims alive in our memory" (Lang, 2007). The National Socialists' attempt to not only physically destroy these human beings, but also erase their existence from the memory of mankind, shall be revoked by the restoration of the victims' names and biographies.

REFERENCES

Angetter D. 1998. Erfassung der von der NS-Justiz in Wien in der Zeit von 1938–1945 Hingerichteten, die als Studienleichen dem anatomischen Institut der Universität Wien zugewiesen wurden. In: Akademischer Senat der Universität Wien, (Editors). Senatsprojekt der Universität Wien: Untersuchungen zur anatomischenWissenschaft in Wien 1938–1945, pp. 81–92. (Unpublished manuscript).

Bever L. 2015. Remains of Holocaust victims found at French forensic institute. *The Washington Post*, July 22. https://www.washingtonpost.com/news/morning-mix/wp/2015/07/22/remains-of-holocaust-victims-used-as-guinea-pigs-found-at-french-forensic-institute/?utm_term=.ea2f09920885 [accessed 31 January 2017]

Blessing T, Wegener A, Koepsell H, Stolberg M. 2012. The Würzburg Anatomical Institute and its supply of corpses (1933–1945). *Ann Anat* 194:281–285.

Cohen J, Werner M. 2009. On medical research and human dignity. *Clin Anat* 22: 161–162.

Czech H. 2002. Der Fall Heinrich Gross. Die wissenschaftliche Verwertung der "Spiegelgrund"-Opfer in Wien. Context XXI, http://www.contextxxi.at/context/content/view/164/93/. [accessed 12 March 2015].

Czech H. 2015. Von der Richtstätte auf den Seziertisch Zur anatomischen Verwertung von NS-Opfern in Wien, Innsbruck und Graz. In: Dokumentationsarchiv des österreichischen Widerstandes (ed.): Jahrbuch 2015: Feindbilder. Wien: Stiftung Dokumentationsarchiv des österreichischen Widerstandes, pp. 141–190.

Forsbach R. 2006. Die Medizinische Fakultät der Universität Bonn im "Dritten Reich." München: R. Oldenbourg Verlag.

Halperin EC. 2007. The poor, the Black, and the marginalized as the source of cadavers in the United States anatomical education. *Clin Anat* 20:489–495.

Hildebrandt S. 2006. How the Pernkopf controversy facilitated a historical and ethical analysis of the anatomical sciences in Austria and Germany: A recommendation for the continued use of the Pernkopf Atlas. *Clin Anat* 19:91–100.

Hildebrandt S. 2008. Capital punishment and anatomy: History and ethics of an ongoing association." *Clin Anat* 21:5–14.

Hildebrandt S. 2011. First Symposium on "Anatomie im Nationalsozialismus" ("Anatomy in National Socialism"), Würzburg, Germany, September 29, 2010. *Clin Anat* 24:97–100.

Hildebrandt S. 2013. Research on bodies of the executed in German anatomy: An accepted method that changed during the Third Reich. Study of anatomical journals from 1924 to 1951. *Clin Anat* 26:304–326.

Hildebrandt S. 2013b. The case of Robert Herrlinger: A unique postwar controversy on the ethics of the anatomical use of bodies of the executed during National Socialism. *Ann Anat* 195:11–24.

Hildebrandt S. 2013c. The women on Stieve's list: Victims of National Socialism whose bodies were used for anatomical research. *Clin Anat* 26:3–21.

Hildebrandt S. 2014a. Stages of transgression: Anatomical research in National Socialism. In: Sheldon R, Benedict S, (Editors). *Human Subjects Research after the Holocaust*, Cham: Springer. pp. 68–85.

Hildebrandt S. 2014b. Current status of identification of victims of the National Socialist regime whose bodies were used for anatomical purposes. *Clin Anat* 27:514–536.

Hildebrandt S. 2016. *The Anatomy of Murder: Ethical Transgressions and Anatomical Science During the Third Reich.* New York: Berghahn Books.

Hildebrandt S, Aumüller g. 2012. Anatomie im Dritten Reich. In: Kühnel W, (Editor). Anatomische *Gesellschaft: Jubiläumsheft 125 Jahre Anatomische Gesellschaft,* Lübeck: Kaiser &Mietzner. pp. 41–48.

Hoffmann A. 1951. Hermann Stieve zu seinem 65. Geburtstag am 22. Mai 1951. Z Mikrosk *Anat Forsch* 57:117–128.

Jones D. 2011. The Anatomy Museum and Mental Illness: The Centrality of Informed Consent. In: Coleborne C, MacKinnon D (Editors). *Exhibiting Madness in Museums: Remembering Psychiatry through Collections and Display*, New York: Routledge. pp. 161–177.

Jones TW, Lachman N, Pawlina W. 2014. Honoring our donors: A survey of memorial ceremonies in Unites States anatomy programs. *Anat Sci Educ* 7:219–223.

Kaelber L. 2014. Am Spiegelgrund, http://www.uvm.edu/~lkaelber/children/amspiegelgrundwien/amspiegelgrundwien.html. [accessed 31 March 2014].

Kenny SC. 2013. The development of medical museums in the antebellum American South: Slave bodies in networks of anatomical exchange. *B Hist Med* 87:32–62.

Kolb S, Weindling P, Roelcke P, Seithe H. 2012. Apologizing for Nazi medicine. A constructive starting point. *Lancet* 380:722–723.

Lang H-J. 2007. *Die Namen der Nummern: Wie es gelang, die 86 Opfer eines NS-Verbrechens zu identifizieren. Überarbeitete Ausgabe.* Frankfurt am Main: S. Fischer Verlag.

Lang H-J. 2013a. Die Namen der Nummern. http://www.die-namen-der-nummern.de/index.php/de/ [accessed 31 January 2017].

Lang H-J. 2013b. August Hirt and "extraordinary opportunities for cadaver delivery" to Anatomical Institutes in National Socialism: A murderous change in paradigm. *Ann Anat* 195:373–380.

Lin SC, Hsu J, Fan VY. 2009. "Silent virtuous teachers": Anatomical dissection in Taiwan. *BMJ* 339: b5001.

Max-Plack-Gesellschaft. 2015. New questions raised by discovery in archives. http://www.mpg.de/9154722/discovery_in_archive. [accessed 6 January 2015].

Mitscherlich A, Mielke F. 1947. *Das Diktat der Menschenverachtung.* Heidelberg: Verlag Lambert Schneider.

Noack T, Heyll U. 2006. Der Streit der Fakultäten. Die medizinische Verwertung der Leichen Hingerichteter im Nationalsozialismus. In: Vögele J, Fangerau H,

Noack T, (Editors). *Geschichte der Medizin- Geschichte in der Medizin,*. Hamburg: Literatur Verlag. pp. 133–142.

Pabst VC, Pabst R. 2006. Danken und Gedenken am Ende des Präparierkurses. Deutsches Ärzteblatt 103(45): A3008–3010.

Peiffer J. 1991. Neuropathology in the Third Reich: Memorial to those victims of National-Socialist atrocities in Germany who were used by medical science. *Brain Pathol* 1:125–131.

Perk W, Desch W. 1974. *Ehrenbuch der Opfer von Berlin-Plötzensee.* Berlin: Verlag das Europäische Buch.

Redies C, Viebig M, Zimmermann S, Fröber R. 2005. Origin of the corpses received by the Anatomical Institute at the University of Jena during the Nazi Regime. *Anat Rec* (Part B: New Anat.) 285B:6–10.

Redies C, Fröber R, Viebig M, Zimmermann S. 2012. Dead Bodies for the Anatomical Institute in the Third Reich: An investigation at the University of Jena. *Ann Anat* 194:298–303.

Richardson R. 2000. *Death, Dissection and the Destitute.* Second edition. Chicago; London: The University of Chicago Press.

Richter G. 2009. Das Arbeitserziehungslager Breitenau (1940–1945). Ein Beitrag zum nationalsozialistischen Lagersystem. Kassel: Verlag Winfried Junior.

Romeis B. 1953. Hermann Stieve. *Anat Anz* 99:401–440.

Sappol M. 2002. *A Traffic of Dead Bodies. Anatomy and Embodied Social Identity in Nineteenth–Century America.* Princeton, NJ: Princeton University Press.

Schmelcher A. 2015. Die Knochen aus dem Uni-Garten. Frankfurter Allgemeine Zeitung, 3 Mai 2015.

Schneider F. 2011. *Psychiatrie im Nationalsozialismus. Erinnerung und Verantwortung.* Berlin: Springer.

Schönhagen B. 1992. Das Gräberfeld X auf dem Tübinger Stadtfriedhof. Die verdrängte "Normalität" nationalsozialistischer Vernichtungspolitik. In: Peiffer J, (Editor). *Menschenverachtung und Opportunismus, Tübingen: Zur Medizin im Dritten Reich* Tübingen: Attempto. pp. 69–92.

Schönhagen B. 1994. Örtlich betäubt: Der öffentliche Umgang mit dem Widerstand gegen das NS-System in Tübingen nach 1945. In: Schnabel T., Hauser-Hauswirth A, (Editors). *Formen des Widerstandes im Südwesten 1933–1945*, Ulm: Süddeutsche Verlagsgesellschaft. pp. 295–309.

Seidelman WE. 1989. In memoriam: Medicine's confrontation with evil. *Hastings Cent Rep* 19(6):5–6.

Seidelman WE. 1996. Nuremberg doctors' trial: Nuremberg lamentation: For the forgotten victims of medical science." *BMJ* 313:1463–1467.

Seidelman WE. 1999. Yizkor: memory and remembrance. Plenary address at: A Long and Winding Road: Research Ethics — the Basics and Beyond", Health Care Ethics Service of St. Bonface General Hospital, Winnipeg, Manitoba, Canada, unpublished manuscript.

Seidelman WE. 2000. Erinnerung, Medizin und Moral: Die Bedeutung der Ausbeutung des menschlichen Körpers im Dritten Reich. In: Gabriel E, Neugebauer W, (Editors). *NS-Euthanasie in Wien*, Wien: Böhlau Verlag. pp. 27–46.

Seidelman WE. 2012. Dissecting the history of anatomy in the Third Reich - 1989–2010: A personal account." *Ann Anat* 194:228–236.

Steinhauser M (Editor). 1987. *Totenbuch Theresienstadt: Damit sie nicht vergessen werden*. Wien: Junius Verlags- und Vertriebsgesellschaft.

Subasinghe SK, Jones DG. 2015. Human body donation programs in Sri Lanka: Buddhist perspectives. *Anat Sci Educ* 8:484–489.

Van Gelder R. 1946. An interview with Erich Maria Remarque. Who outlines his rules for handling luck, loneliness and refugee living. January 27, 1946. In: van Gelder R, (Editor). *Writers and Writing*, New York: Charles Scribner's Sons. pp. 377–381.

Weindling PJ. 2010. Psychiatrische Opfer von Humanexperimenten im Nationalsozialismus. "Jeder Mensch hat einen Namen". *Psychiatrie* 7(4): 255–260.

Weindling PJ. 2012. "Cleansing" anatomical collections: The politics of removing specimens from anatomical collections 1988–1992. *Ann Anat* 194:237–242.

Weindling PJ. 2013. From scientific object to commemorated victim: The children of the Spiegelgrund. *Hist Phil Life Sci* 35:415–430.

Winkelmann A. 2007. Die menschliche Leiche in der heutigen Anatomie. In: Graumann S, Grüber K, (Editor). *Grenzen des Lebens. Beiträge aus dem Institut Mensch, Ethik und Wissenschaft*, Band 5, Berlin: LIT Verlag Dr. W. Hopf. pp. 62–74.

Winkelmann A, Güldner FH. 2004. Cadavers as teachers: The dissecting room experience in Thailand. *BMJ* 329:1455–1457.

Winkelmann A, Schagen U. 2009. Hermann Stieve's clinical-anatomical research on executed women during the "Third Reich". *Clin Anat* 22:163–171.

5

COMPLEXITIES AND REMEDIES
OF UNKNOWN-PROVENANCE OSTEOLOGY[a,b]

Carl N. Stephan*,§, Jodi M. Caple*, Andrew Veprek[†],
Emma Sievwright*, Vaughan Kippers[‡], Steve Moss[†]
and Wesley Fisk[†]

*Laboratory for Human Craniofacial and Skeletal Identification (HuCS-ID Lab),
School of Biomedical Sciences, The University of Queensland, Brisbane, Australia,
[†]Gross Anatomy Facility (GAF), School of Biomedical Sciences,
The University of Queensland, Brisbane, Australia
[‡]School of Biomedical Sciences, The University of Queensland, Brisbane, Australia
[§] c.stephan@uq.edu.au

ABSTRACT

Unknown-provenance osteology is the rule rather than the exception at most well-established university anatomy laboratories. This circumstance results from the difficulties associated with processing skeletons and consequent historical reliance on external skeleton supply — one common source is the Calcutta bone trade that operated at its peak from 1930 to 1985. This trade shipped tens-of-thousands of skeletons across the

[a]Portions of this work have previously been presented at the: 29th Annual Meeting of the Australasian Society for Human Biology, Brisbane, Australia, 2015; and 12th Annual Meeting of the Australian & New Zealand Association of Clinical Anatomists, Adelaide, Australia, 2015.
[b]The views and opinions contained herein are solely those of the authors and are not to be construed as official, or as views of the School of Biomedical Sciences or The University of Queensland.

globe *each year* (notably without accompanying consents), potentially exceeding 2 million skeletons in total, dwarfing numbers associated with body-snatching that gave rise to the Anatomy Act in Britain. These events are extraordinary in anatomy's history, yet it receives next to no acknowledgement in the scientific literature and countermeasures to address this situation have drawn little attention. Herein, we review articles from the popular press to provide an overarching account of the Calcutta bone trade and supplement this review with osteology from The University of Queensland's (UQ) anatomy rooms. We outline novel and positive steps taken at The School of Biomedical Sciences to recognize and decommission unknown-provenance osteology (in all probability derived from the Calcutta bone trade) to a memorial assemblage — an action made possible via the simultaneous establishment of a new osteological collection of indefinite body donors. While the memorial assemblage cannot undo events of the past, this action awards these unknown and potentially misappropriated individuals levels of respect not previously afforded. This approach is not only favourable from an ethical standpoint, but it also offers substantially improved osteological morphologies to benefit student learning.

INTRODUCTION

The authentic osteology housed at most University anatomy laboratories around the world includes individuals of unknown identity, in many cases acquired prior to 1985 and without accompanying consents of the subjects concerned (Hefner *et al.*, 2016). Consequently, the osteology is of unknown provenance and in some cases may have been taken without next-of-kin permission. It is a widely known though little discussed fact that the vast majority of acquired skeletons derive from a bone trade out of India from 1930 to 1985 (Anonymous, 1976; Hefner *et al.*, 2016).

The scale and extent of this trade, up to 65,000 individuals per year (Anonymous, 1989) for 55 years, is staggering and to fully appreciate the circumstances that spawned this 'event' one must understand five key factors: (1) osteology is a core component of human anatomy; (2) anatomy has a historic culture of body-snatching associated with cadaver acquisition; (3) the first anatomy act did not prohibit the acquisition of bodies without accompanying consents (especially of skeletons from abroad);

(4) the demand for skeletons far exceeded that of cadavers for dissection; and (5) modern-day anatomy acts are, in comparison to skeleton acquisition practices, very recent. We briefly outline each of these items below as necessary for a full and complete appreciation of the Calcutta bone trade, which is thereafter reviewed by drawing upon accounts only available in the popular press.

Osteology is Core to Human Anatomy

The key role osteology plays in human anatomy is classically illustrated by its review ahead of other body systems in the opening chapters to the iconic and authoritative first edition of *Anatomy, Descriptive and Surgical* by Gray (1858). Lockhart also captures its essence when he states "[bones] carry the burden of the subject; the whole story literally hangs upon them and they will teach two-thirds of anatomy" (Lockhart, 1927). It is not surprising then that anatomists across history have sought skeletons to study and train their students.

This stretches far back, for example, to the father of modern anatomy, Andreas Vesalius. Vesalius reportedly took several skeletons including one from the gallows outside the city walls of Louvain (Corner, 1930). He also stole bones from the Cemetery of the Innocents (plague victims) and from criminals hung at Montfaucon (Saunders and O'Malley, 1993; Nuland, 1995). This osteology not only enabled the first illustration of the human sphenoid bone (Singer, 1957), but it was osteology that further inspired Vesalius' most influential anatomical work *De humani corporis fabrica libri septem*, for while Vesalius was articulating a skeleton of a priest with access to that of a babary ape, he realized that Galen must have heavily relied on animal, not human, anatomy (Shapiro, 1972). As an example, the anapophysis, a posterior articular processes on the neural arches of the vertebra, described by Galen with respect to humans, was only present so far as Vesalius could see, in the macaque thereby elucidating Galen's vital mistake (Shapiro, 1972). This prompted Vesalius to correct more than 200 of Galen's errors in human anatomy paving the foundations for modern scientific human anatomy practice, which was based on the Renaissance theme of seeing for one's self in real human subjects (Franklin, 1949; Singer, 1957; Russell, 1973; French, 1997).

The Rise of Body-Snatching

The utility of scientific observation established by Vesalius and others during the Renaissance found popularity for medical training in France during the eighteenth century (Russell, 1973). Its utility was further recognized by the surgical fraternity in the United Kingdom to legitimize their craft as a science-based pursuit, elevating surgery to loftier social standings.

The Hunter brothers played an important role in that effort (Moore, 2005) and eventually surgery took on similar esteem awarded to general medicine. This saw surgeons unshackling themselves from the Barbers Guild and establishing the Royal College (of Surgeons) by official charter in 1745 (Major, 1954; Singer and Underwood, 1962; Russell, 1973; Robinson, 1984; Himmerlmann, 2007). The embedding of observation-based dissection in medicine drove a demand for cadavers that, in the United Kingdom, infamously resulted in a cadaver shortfall (Russell, 1973).

Up until 1752, the only bodies made available for anatomical purposes in Britain were 1–6 criminals per year, which was changed to all executed criminals in 1752 with the passage of the Murder Act (Guttmacher, 1935; Richardson, 2000). However, supply did not keep pace with demand and subsequently the underground practice of grave robbing or body-snatching was cultivated (Ball, 1928; Guttmacher, 1935; Cole, 1964; Ross and Ross, 1979; Shultz, 1992; Tward and Peterson, 2002). Not too infrequently, this body-snatching was conducted by the anatomists, physicians, surgeons and/or their students leading to a moral controversy readily pointed out in the social science literature (Richardson, 2000; MacDonald, 2005, 2006, 2010; Hurren, 2014).

Acknowledgments by the first six presidents of the New York Academy of Medicine that they had participated in body-snatching provides an indication of how frequent the practice was, not only within the UK where the cadaver shortage first manifested, but also in other countries that followed similar suit (Shultz, 1992). Ultimately and perhaps inevitably securing bodies for payment culminated in homicides to acquire saleable material for dissection (Anonymous, 1829) — the most famous of which are the murders committed by Burke and Hare (Anonymous, 1829; Ball, 1928; Hunter, 1931; Cole, 1964), but there are a number of other cases, e.g., a boy and a woman drowned after suffering a broken neck by John Bishop and Thomas Williams around 1831 (Shultz, 1992).

The Anatomy Act of 1832

Tensions ran high between anatomists and members of the general public, who not infrequently rioted at houses of the former when they were found to be endorsing body snatching (Ball, 1928; Cole, 1964; Shultz, 1992). The anatomists began paying guards to hold watch during their dissection proceedings (Hunter, 1931), while the public took their own measures to protect the dead: placing guards at cemeteries; using mortsafes and iron coffins to protect their contents; setting trip wires, spring guns, booby traps and even explosive devices to deter grave robbers; and preparing bodies for burial in locked decomposition crypts (Ball, 1928).

While body-snatching and homicides might seem reason enough to implement an act that addresses body supply for medical teaching, this was not the primary driver. Prior to the Burke and Hare murders, two cases emerged where anatomists/physicians were prosecuted and convicted for having caused a body to be disinterred, leaving many practicing anatomists liable to prosecution (Richardson, 2000). Under the pressure from the medical community to change the laws, the British parliament established a Select Committee to investigate the matter in 1828 (Guttmacher, 1935; Richardson, 2000).

After interviewing forty people to produce a report (25 persons of which were members of the medical profession and 17 of whom were surgeon-anatomists), the Select Committee concluded that a paramount need existed to study anatomy by dissection and that unclaimed bodies should be used as this source (Guttmacher, 1935; Richardson, 2000).

An anatomy bill was introduced into the House of Commons in 1829; however, this bill was defeated in the House of Lords (Guttmacher, 1935; Richardson, 2000). A second revised bill was submitted and passed in the House of Lords in 1832, ten days after the execution of Bishop and Williams (Guttmacher, 1935; Richardson, 2000). The 1832 British Anatomy Act created a supervisory board of inspectors and abandoned the compulsory dissection of executed criminals in favor of unclaimed bodies — which essentially meant the poor (French, 1997; Richardson, 2000; Gill *et al.*, 2001). It also provided provisions for persons to bequeath their bodies for dissection (Richardson, 2000).

Notably, the Anatomy Act of 1832 did not prohibit grave robbing, rather it provided provisions (the use of unclaimed bodies) to circumvent its demand. While insufficient early on, this provision more-or-less fulfilled its objective of supply after the advent of embalming, which enabled cadavers to be used over much longer time frames (MacDonald, 2010). It is worth bearing in mind that the extent of cadavers reportedly sold for dissection across the entire Victorian era in London is estimated to total 60,000 to 125,000 (Hurren, 2014).

Skeleton Demand

The demand for cadavers to dissect was only part of the overall demand for human bodies. Medical students were often expected to possess their own authentic human skeleton for private studies (preferably white pristinely colored skeletons according to Hunter (Moore, 2005)), creating a skeleton demand that far outstripped whole cadavers. At some institutions, this preference existed as recently as the 1970's, for example it was a requirement in The University of Queensland's medical program at that time (Fig. 1). Given the collective numbers of medical students across the developed world at the turn of the twentieth century, the demand for skeletons was large and led to commercial opportunities for skeleton supply and bone import businesses (unchecked by the Anatomy Act of 1832 in Britain).

The result was up to 65,000 skeletons shipped annually from India at the height of the trade (Anonymous, 1989). Almost all skeletons came from corpses that were stolen from graves, procured from rivers, or diverted from crematoriums largely without next-of-kin permission. As far as the numbers go, the entire cadaver demand from Victorian era Britain equates to only a few years of osteological exports from India.

The First Traces of the Calcutta Bone Trade

The first report of skeletons being exported from India comes from 1944 (Fig. 2), in the middle of World War II and following the Bengal Famine of 1943 (Litten, 1944). Within one year, this devastating famine reportedly claimed the lives of between 1.5 to 4 million people — *Life Magazine* estimated 50,000 people per week at the time (Fisher, 1943). Initiated by

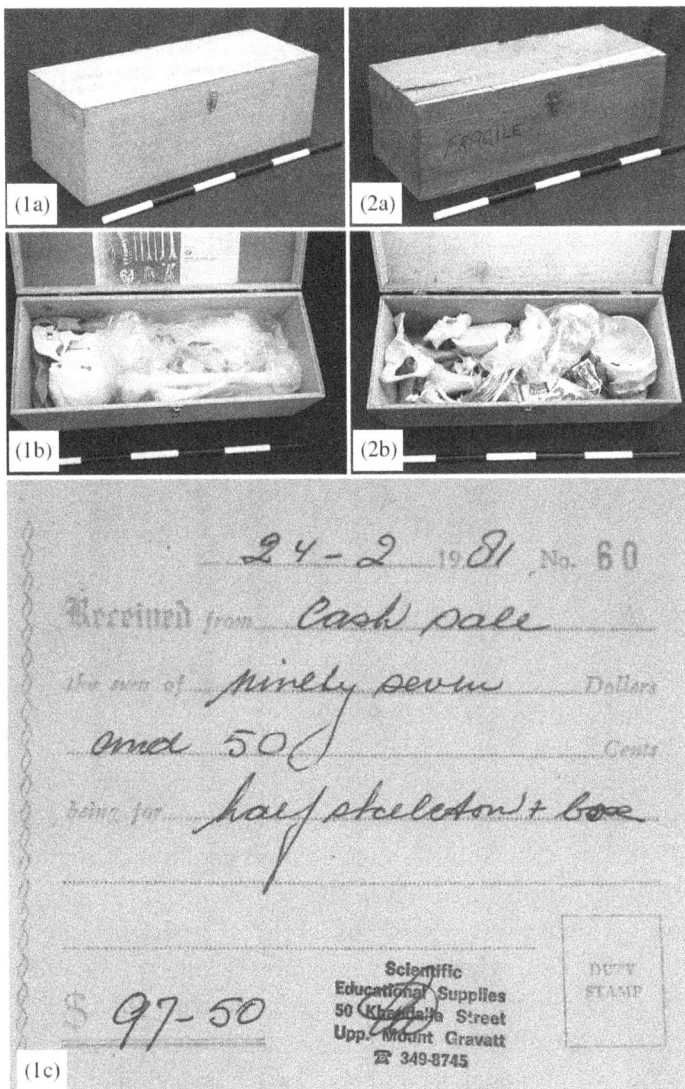

Figure 1. Two examples of typical South-Asian half-skeletons used by anatomy students in Australia for medical study: (1a & 2a) original storage boxes; (1b & 2b) skeletons contained therein; and (1c) original receipt of the skeleton for (1a & 1b) from medical suppliers that are no longer trading. Of note, the sale date in this instance falls after the introduction of the Queensland Transplantation and Anatomy Act, 1979. Instances of a two-year lag between Anatomy Act introduction and cessation of sales have also been reported elsewhere around the country — see Russell (2011) with regards to Victoria's Human Tissue Act of 1982. Scale bar intervals = 100 mm.

Figure 2. *Life Magazine* cover, and article, that documents the bone export trade from India in 1943 following the West Bengal Famine.

a cyclone in 1942, the famine was partly fueled by Winston Churchill's scorched earth policy in South Asia: food diversion from the famine struck area to slow any Japanese advance. This provided ample opportunity for a profiteering Indian to expand his skeleton business (Litten, 1944; Banerjie, 1985) and in view of the profitability, others followed with 13 skeleton exporters operating out of Calcutta (Banerjie, 1985; Spink, 1992; Majumdar, 2006). Collectively these bone merchants reportedly employed around 300 staff — enough to start their own association of bone exporters called "The Association of Exporters of Anatomical Specimens" (Banerjie, 1985).

Reknas Ltd. was one of the well-known Calcutta bone merchants, supplying material to Adam Rouilly in the UK from the 1930's through to the mid 1980's (http://www.adam-rouilly.co.uk/content/history.aspx; last accessed 7 Aug 2016). Indeed, this timing corresponds closely to the establishment of the Anatomy Laboratories at The University of Queensland

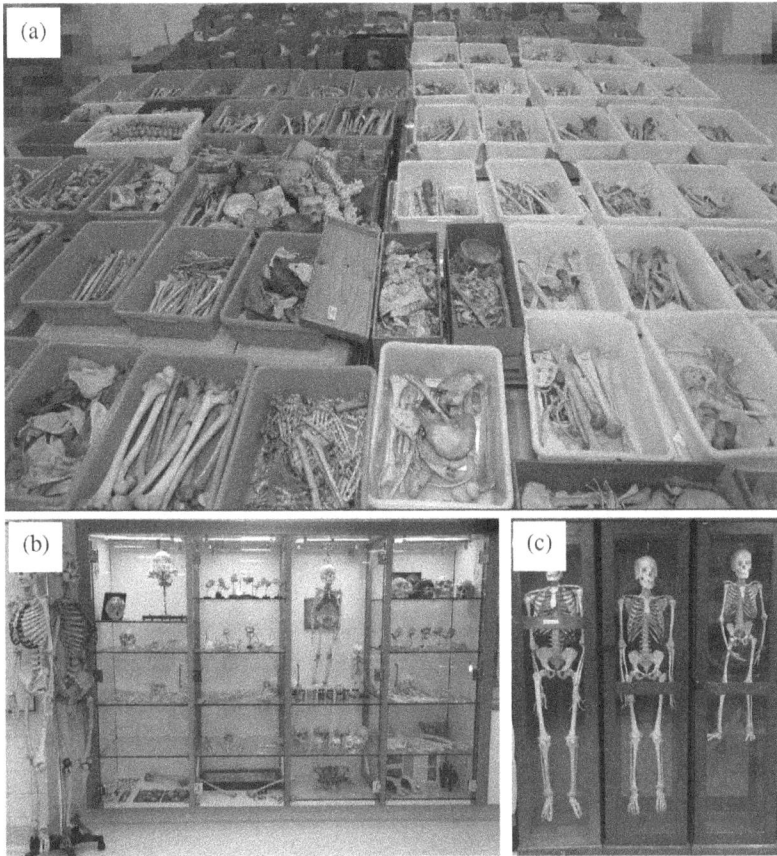

Figure 3. The present-day osteological collection at UQ: (a) disarticulated elements for student study; (b) cabinet holding juvenile and other delicate osteology plus two additional articulated skeletons; and (c) three other articulated skeletons.

(see above) and its osteological collection (Fig. 3). Materials marked with Adam Rouilly's insignia, consistent with these South-Asian imports, are present within the collection (Fig. 4). Here it must be emphasized that this collection is by no means a typical for anatomy departments, nor is the UQ collection large by comparison to other collections in Australasia.

In many respects the import of skeletons from India to Britain, for further distribution, was convenient: India was British governed at the time facilitating imports and India was largely Hindu so the body counted for much less after the soul was thought to have departed (the body was

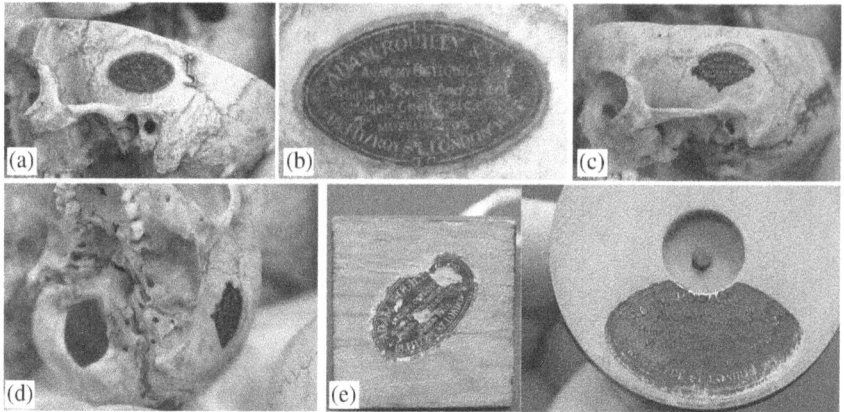

Figure 4.   Osteology bearing the Adam Rouilly seal: (a) adult skull; (b) enlarged view of label in (a); (c) juvenile skull (lateral view); (d) juvenile skull (oblique view) showing open spheno-occiptial synchondrosis; and (e) underside of two temporal bone stands.

replaced with each reincarnation). Consequently, opportunity existed for skeletons to be acquired without adding substantial strain to western cadaver supplies, these skeletons could be acquired *en masse* with little risk of Western public backlash, and benefits included training new doctors to assist the war effort (Litten, 1944).

Mass Skeleton Export

Prior to 1985, Calcutta exporters reportedly traded $1.42 million worth of skeletons to 23 countries each year (Anonymous, 1981; AFP, 1985; Fineman, 1991b) — some reports go as high as $5M or $6.7M per year (Anonymous, 1985a; Corrales, 1987). The USA reportedly became the biggest customer (AFP, 1985; Fineman, 1991b) importing $622,000 worth of material in 1984 (Corrales, 1987). Allegedly, it was routine for 10,000–15,000 skeletons and 50,000 skulls to be shipped annually from Calcutta (McConnaughey, 1986; Anonymous, 1989). Demand was such that skeletal processing facilities became family run businesses (Banerjie, 1985; Fineman, 1991a; Chatterjee and Sarkar, 2007), many operating for multiple generations prior to the permanent export ban of 1985 (see Section on "Skeleton Export Bans"). Universities and high-schools were such regular customers that the bone traders reported receipt of Christmas cards and letters annually from these customers (Fineman, 1991a, 1991b).

Skeleton Acquisition

So far as we are able to determine, detailed accounts of South-Asian skeletal acquisition methods are absent from the written pre-1985 records; however, insight can surely be gleaned from bone merchants that operated past the 1985 export ban and which, therefore, came under police and media scrutiny. These include reports from Mukti Biswas, a 62 year old Bengali who operated a Jogeshwar burning ghat on the banks of Bhagirathi River and who was raided by the police in 2007 (Anonymous, 2007a).

The Bhagirathi river (also known as the river Ganga) is one of two head streams of the Ganges River, the latter of which is regarded to be especially holy and sacred by Hindus. Day and night, literally hundreds of corpses are placed in the Ganges either as cremated ashes from burning ghats along the banks, or for poorer families whole bodies released from further upstream. The Ganges receives the most bodies because it is deemed to be the most sacred river, with as many as 100,000 intact corpses thought to be released into the Ganges every year (McBride, 2014). Even today, corpses floating in the Ganges are common, as evidenced by the simultaneous banking of 100 bodies in 2015 that generated media headlines (Chaturvedi, 2015).

According to Biswas, he fished corpses from the Ganga to obtain skeletons (Anonymous, 2007a, 2007b; Chatterjee and Sarkar, 2007). Locals explain that these corpses were subsequently submerged in the river by being tied to bamboo stumps, so that local wildlife would remove the decomposing flesh (Anonymous, 2007a). The remnants of the body were then placed in earthen pots with caustic chemicals and boiled (Anonymous, 2007a, 2007b; Chatterjee and Sarkar, 2007). The bones were retrieved, dried, packed and sold (Anonymous, 2007a). Neighbors, from up to 15 km from Biswas's processing plant, reportedly complained of the smell to local authorities on a semi-regular basis (Anonymous, 2007b; Chatterjee and Sarkar, 2007). In addition to 'fishing', the Burdwan superintendent of police reported that the raids on Biswas' processing plant were motivated by accounts that Biswas was also stealing corpses from local cemeteries (Chatterjee and Sarkar, 2007).

Both river 'fishing' and cemetery robbing are confirmed by interviews made by Carney (2011) with one of Biswas's workers. According to this source, corpses were wrapped in netting, anchored in the river where the bodies were reduced to bones in a week (Anonymous, 2014). The bones were then scrubbed and boiled in a cauldron of water and caustic soda (Anonymous, 2014). This process reportedly left the bone with a yellow tint, so they were subjected to sun bleaching for a further week to turn them white, before being soaked in hydrochloric acid as a final processing step (Anonymous, 2014).

The robbing of skeletons from cemeteries has also been an ongoing issue in India (Anonymous, 1985a), so much so, that locals have requested police guards on gates (Anonymous, 2014) and/or established their own security patrols in the form of child guards (Carney, 2011). There are also reports of bone merchants purchasing skeletons from the living ahead of their deaths to guarantee a constant chain of supply (Spink, 1992).

Skeleton Export Bans

Across the twentieth century, the Indian government implemented three skeleton export bans in total (Banerjie, 1985; Putka, 1986). The first two bans, one in 1952 (Banerjie, 1985) and one in 1976 (Putka, 1986; Fineman, 1991a) were very short lived — a few months in the case of the first and approximately one year in the case of the second. Only the third and final ban, established in 1985, has persisted.

After the second ban fell, a legacy was left that police certificates were to be issued to bone merchants as evidence that the bodies they held had been collected legally, but the police allegedly began accepting bribes for the paperwork (Fineman, 1991a). This lead to an outcry, with tensions running so high that after reports of 1,500 children being kidnapped from Calcutta streets for their skulls (AFP, 1985), an Australian tourist was mistaken for a body-snatcher and beaten and burnt to death in Shrinagar by a 4,000 strong mob (Anonymous, 1985b, 1985c; Fineman, 1985). The 1985 ban thereby coincided with a peak of discontent, and has remained in place to this day, along with local statutes against grave desecration

(Anonymous, 2014). In the first year following the ban (1986), the export market was reduced from over a million dollar export trade to reportedly $30,589 (Corrales, 1987). There have been remittent and ongoing reports of skeleton trading since the 1985 ban (Anonymous, 2007a, Chatterjee and Sarkar, 2007; Anadrabi, 2009; Gupta, 2011), but the bone trade seems set to almost entirely disappear with local Human Tissue Acts of other countries that make the sale of any human tissues, including extracellular bone matrix, illegal.

Since the 1985 prohibition, bone merchants, medical suppliers and anatomists have largely switched to plastic replicas, begrudgingly in the view of some (Anonymous, 1976, 2010; Banerjie, 1985; McConnaughey, 1986; Putka, 1986; Corrales, 1987, 2014), and the price of authentic bone specimens has soared (Fig. 5). To put these prices in perspective, during the UK's cadaver shortfall the price of a corpse increased 'only' 16-fold, from 1 guinea to 16 guineas across the thirty years in the lead up to the Select Committee's Report (The Select Committee on Anatomy, 1828). In contrast, prices for skeletons have risen 26-fold from 1944 ($US25) (Litten, 1944) to 1989 (US$650) (Anonymous, 1989) and now routinely

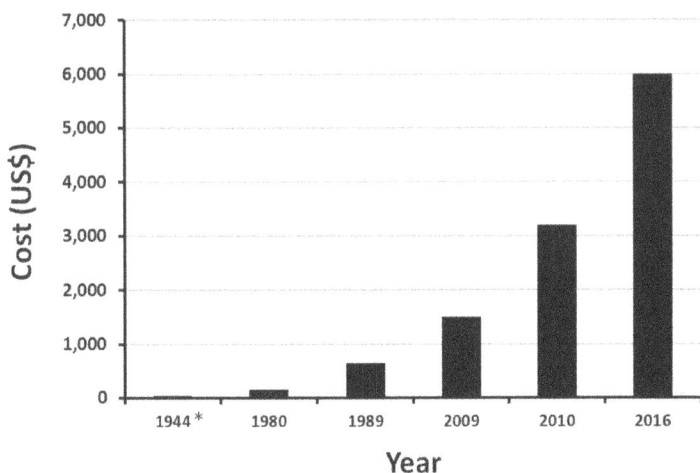

Figure 5. Increase in US market prices for authentic human skeletons since 1944. Note the 1944 price has been doubled to US$50 to visually register on the graph. Data sourced from Litten (1944), Finneman (1980), Anonymous Article in the Manila Standard (1989), Andrabi (2009), Carney (2010) and an USA retailing website (2016).

fetch as much as 240x their original price in the USA. Note here that on occasion, some exceptional prices have been paid for single skeletons throughout history — one example being John Hunter's purchase of Charles Bryne's skeleton ("The Irish Giant") in 1783 for £500 (Guttmacher, 1935).

A MODERN-DAY WINDOW INTO THE CALCUTTA BONE TRADE

First established as an authorized anatomical facility in 1927 under the Queensland Medical Act of 1925, The University of Queensland's (UQ) School of Biomedical Sciences Gross Anatomy Facility houses a skeletal collection (Fig. 3) that came into existence during the prime of the South-Asian exports. The collection holds many of the hallmarks of South-Asian material, some of it traceable to medical suppliers known to be dealing in South-Asian material (Fig. 4) and it provides one window of insight into the Calcutta bone trade. For example, skeletons commonly comprise young individuals, which are commingled (but with parts often matched according to gross size). The skeletons in many cases are over-processed (hot water macerated and/or bleached to chalky white consistency), and sharp trauma to bones around articulating surfaces in some instances record unskilled disarticulation practice.

A skull with an affixed Adam Rouilly label in Fig. 4d is consistent with the young profile of much of the South Asian material and the affixed label provides further ties to the Calcutta bone trade. This skull exhibits a partially fused pars basilaris and pars lateralis and open spheno-occipital synchondrosis indicative of an individual below the age of 7 years (Scheuer *et al.*, 2000). Young ages (< 22 years) are also indicated by unfused iliac crests that are not uncommon in the collection (Fig. 6), as are the unfused ventral margins of S1 and S2 segments. Two 'half' skeleton examples from Fig.1 further exemplify characters common to the South Asian skeleton trade, including their commingled state, pieced together from different individuals to give the appearance of one person (Figs. 7 and 8).

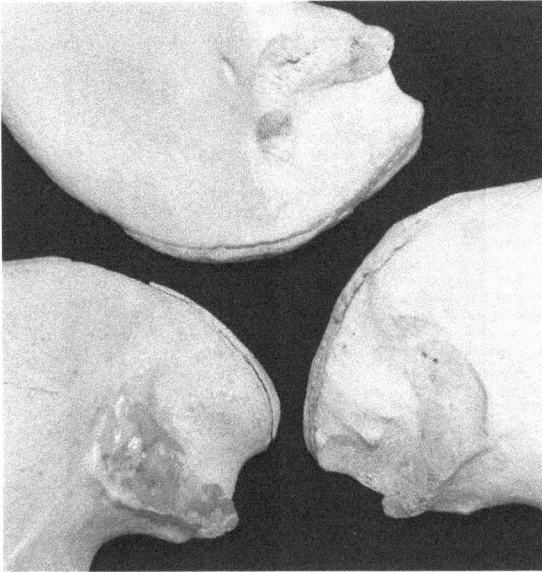

Figure 6. Examples of unfused iliac crests from different individuals in the UQ Osteological Collection. (Glue obscures either part or all of the auricular surfaces on two os coxae depicted here.)

The first osteological set (Fig. 9) discussed here is clearly commingled, possessing a much younger infra-cranial skeleton (Fig. 9a–d) that contrasts with the much older adult skull exhibiting tooth loss (Fig. 9e) — commingling is also confirmed by non-articulation of the atlas with the occipital condyles (Fig. 10), and by mismatched articulations between C2-5, sacrum and coccyx and additional coccygeal segments (Fig. 10). This half skeleton is comprised of at least two individuals, but could be as many as five. Such commingling is consistent with reports that avoiding commingling was difficult at the time of skeletal processing and that bone merchant staff received bonus pay in an attempt to keep skeletons together and attract higher prices (Carney, 2011). The left shoulder of this skeleton also illustrates cut marks to the inferior scapular spine, glenoid fossa and posterior humeral head (Fig. 11) indicative of unskilled disarticulation.

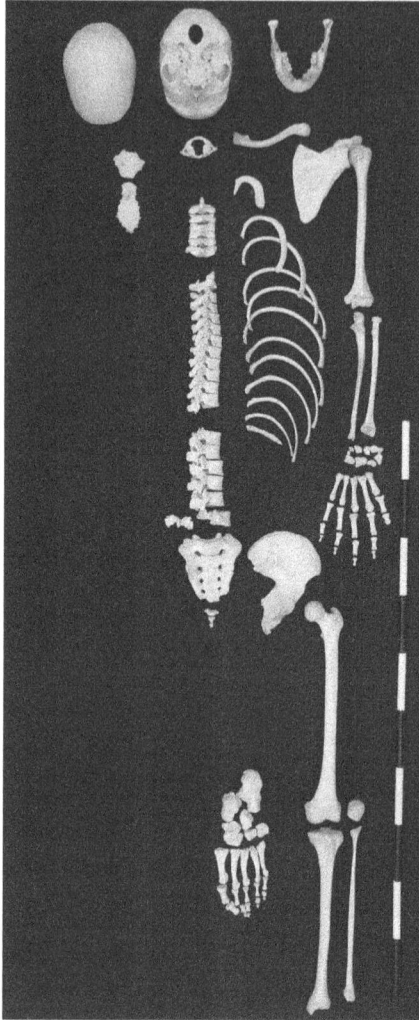

Figure 7.   Skeletal layout of osteology contained in first box from Fig. 1 (1a & 1b; with receipt of purchase); here on termed Skeleton1. Scale bands are 100 mm.

The second 'half' skeleton (Fig. 8), presents a similar picture. This skeleton is classically female judging by the pelvis (e.g., wide sciatic notch), but male according to the skull (e.g., large brow ridges; Fig. 8); indicative of commingling. Examination of the atlanto-occiptal joint

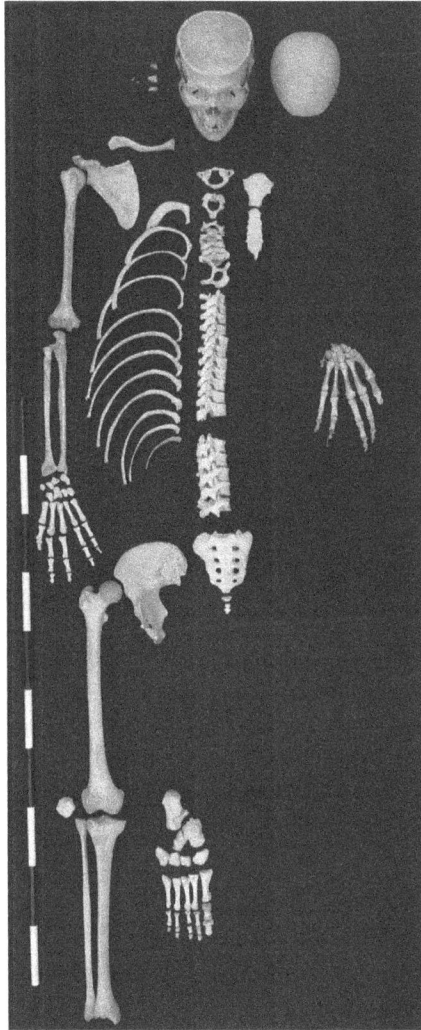

Figure 8. Skeletal layout of osteology contained in the second box from Fig. 1 (2a & 2b); here on termed Skeleton2. Scale bands are 100 mm.

reveals a mismatch in articulation with the occipital facets on the atlas too small and shallow for the larger occipital condyles (Fig. 12). Moreover, this skeleton is accompanied by an additional right hand from a much larger individual (Fig. 8). Consequently, the skeleton consists of elements from at least two, possibly three individuals.

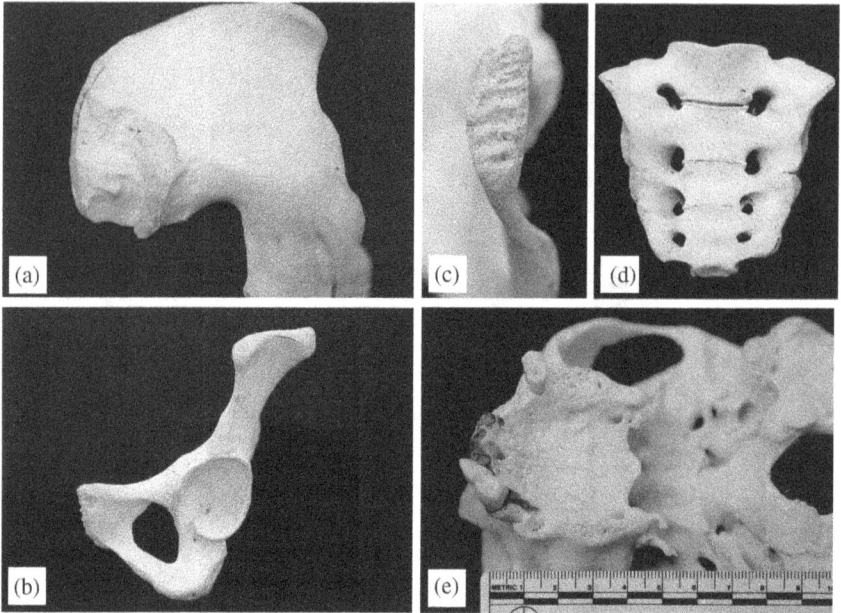

Figure 9.   General biological character of Skeleton1: (a) female indicated by the wide sciatic notch and young age (still fusing iliac crest); (b) female os coxae morphology from anterior view and again open iliac crest; (c) billowy and young appearance of pubic symphysis; (d) young sacrum indicated by still open ventral sacral junctions; and (e) skull consistent with an older individual judging by adult morphology, tooth wear, loss and closure of the spheno-occipital synchondrosis.

The profile of these example skeletons is representative of others in the set and it precisely aligns with South-Asian skeleton trade: incomplete skeletons have been pieced together to represent "whole" half skeletons and are thereby commingled. This commingling is consistent with reports of Indian skeletisation techniques that left ample opportunity for segment loss and/or mixing (bodies being submerged in nets to decompose in the river/bodies stashed in holes along the river bank to decompose/skeletons haphazardly boiled down in pots/skeletons laid out on rooves to sun-bleach (Anonymous, 2007a). Skulls have classic Caucasoid appearance and infra-cranial skeletons are often of younger

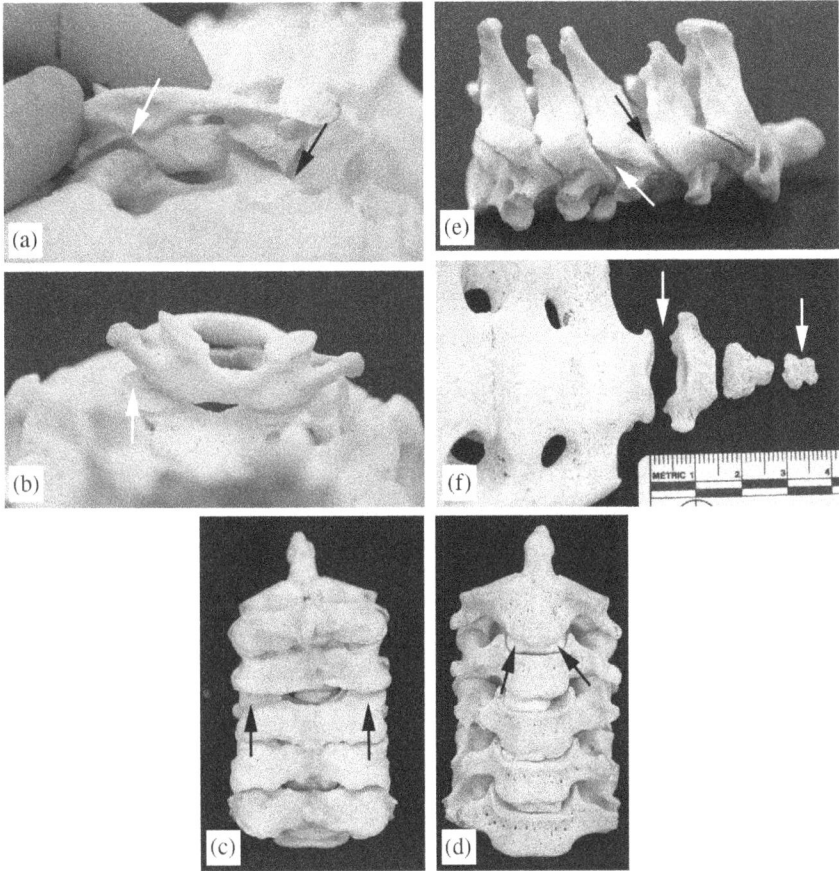

Figure 10. Commingling in osteological set of 'Skeleton1'. (a) Incongruent atlanto-occipital articulations — atlas not wide enough (black arrow), and facets too deep (white arrow). (b) different view of (a) showing joint alignment on the left side and mismatch on the right (white arrow); (c) mismatch in alignment and size of C3-4 zygapophyseal joint surfaces; (d) incongruent shape of C2 body with C3 vertebra; (e) incongruent C4 zygapophyseal joints for this vertebral series; and (f) dubious S5-C1 articulation and commingled coccyx — segments 1-3 belong, but conjoined 3-4 do not.

individuals. Over processing of bones reducing them to chalky white appearance is also evident in some cases, and consistent with Indian maceration techniques that often included concoctions of caustic soda, acid and/or thermal heat.

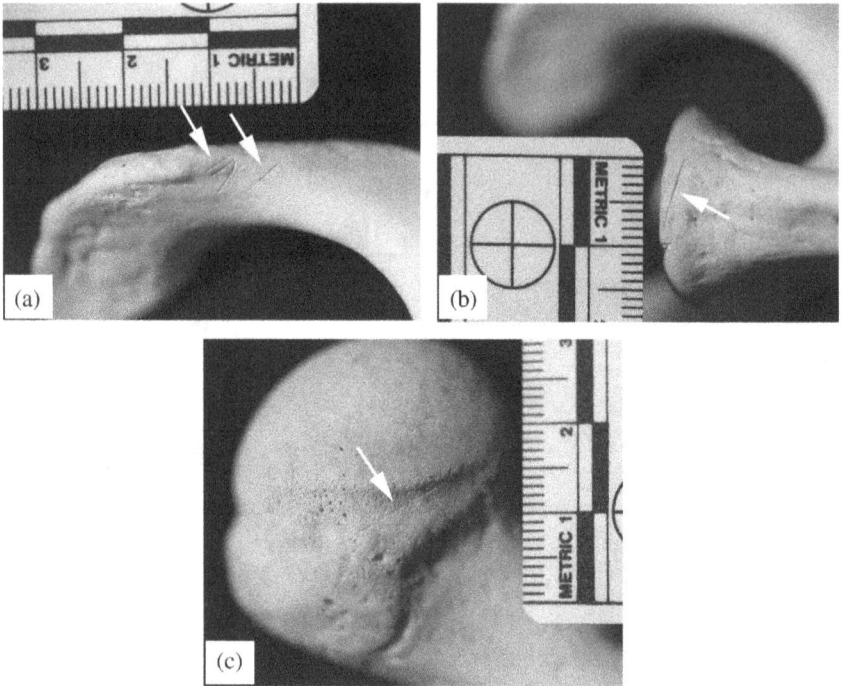

Figure 11.   Sharp force trauma (white arrows) to the shoulder of Skeleton1: (a) inferior margin of the scapular spine; (b) inferior glenoid fossa; and (c) posterior humeral neck.

THE LINGERING DILEMMAS FROM THE CALCUTTA BONE TRADE

This unknown provenance material, much of it likely derived from the Calcutta bone trade, presents a number of problems:

— *There is no accompanying consent given either by the subjects or by their next-of-kin*; which breaches modern ethical standards. The fact that body-snatched material was acquired before legislation legally prevented its procurement does not excuse the embedded ethical complexities or its continued use, and certainly solutions to these conundrums are well overdue. The acquisition of the South-Asian osteology under for-profit sale adds yet another layer of complexity that runs counter to established ethical standards described by various, more

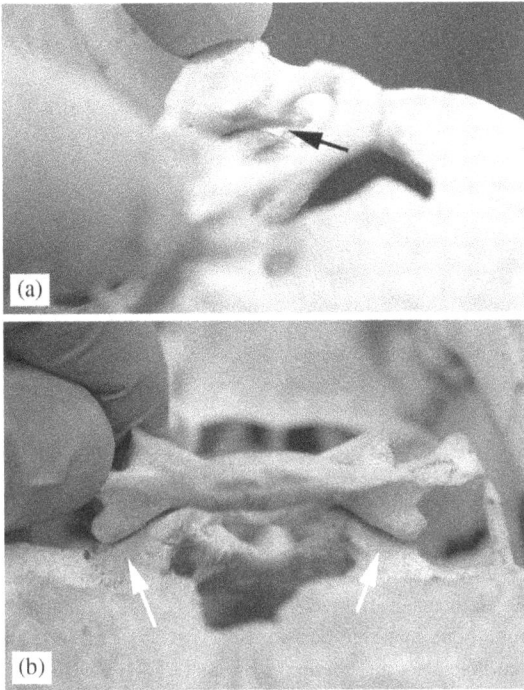

Figure 12. Commingling in osteological set of Skeleton2: (a) incongruent atlanto-occipital articulation; and (b) posterior misalignment after aligning ventral articular surfaces.

recently introduced, Human Tissue Acts (e.g., Queensland Government, 2016).

— *The osteology is not representative.* South-Asian skeletons are commonly incomplete, commingled, and young. The osteology thereby presents a misleading picture of that belonging to a single person and fails to portray adult anatomy described in introductory anatomy textbooks, such as that for muscle attachments which are difficult to see on the bones of younger individuals. Moreover, the over-processing conducted by the Indian bone merchants producing a chalky consistency to the bone, results in a loss of surface detail that further decreases the educational value of the material.

— *The material is not sustainable.* Handling of material by students inevitably results in wear over time, however, the Indian export ban

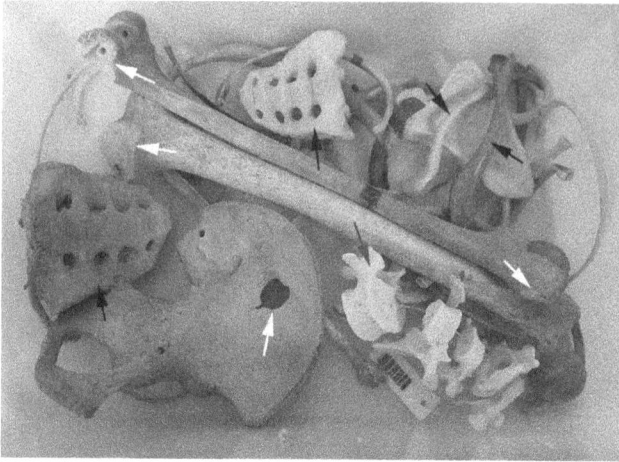

Figure 13.   One of UQ's current osteological teaching sets used by undergraduate students. Note: the damage to skeletal elements from years of use (white arrows); commingling - replicate sacra and clavicles of vastly different size (black arrows); and some mixing with plastic elements (vertebrae — grey arrow). Also note again young ages of some of the authentic bone material, i.e., unfused ventral sacral junctions in the smaller sacrum, and unfused iliac crest margin on the os coxa.

prevents the replacement (Putka, 1986) as does prohibition by local Anatomy Acts that prevent new purchases (Queensland Government, 2016). Consequently, the deteriorated osteology is posed for further deterioration.

POSITIVE NEW STEPS

These hurdles may be surpassed by retiring the unknown provenance osteology to a memorial assemblage for future possible repatriation, and its replacement with authentic skeletons from indefinite body donors who have bequeathed their bodies for anatomical study. This is one of our current undertakings at UQ.

Across the globe, there have only ever been a small handful of university departments that have regularly produced their own skeletons, and those for anatomical teaching in anatomical departments are even sparser, especially when presently active skeletisation programs are considered (Table 1). This is due in part to two perceptions: (1) that skeletisation leaves nothing of the rest of the body for remains to be returned to families under bequests (Putka, 1986); and (2) that skeletal processing is a messy and offensive business that precludes a dignified undertaking (Anonymous Article, 1976; McMaster University of Health Sciences, 2009). Both of these assertions are invalid.

First, body donors may bequeath their bodies under indefinite intent. Second, if conducted in a properly organized, systematic, expeditious and tightly controlled manner, skeletisation can be conducted respectfully without offensive odor. Consequently, there is no reason why bodies from affluent societies cannot be used to support anatomical training, rather than undertaking the questionable importation of skeletons from much poorer societies.

Skeletisation Process

At UQ skeletisation is undertaken by harnessing the best of a natural process, that is *Dermestes maculatus* activity, producing a method that is fast, natural and respectful and which produces osteology of exceedingly high standards, including retainment of all bones down to the very small sesamoids of the pollex and tracking of all elements to individual, side and number (e.g., for similar body parts such as digits). As explained in step-by-step detail elsewhere (Stephan *et al.*, 2016), when conducted with a rigorous quality assurance program, this procedure enables the three intermediate stages of decomposition (bloat, active decay, advanced decay (Payne, 1965)) to be avoided enabling the skeletisation process to essentially jump from the first initial stage (fresh) to the fifth and final stage (dry remains) within a matter of hours (commonly <12h). This avoids the odorous components associated with maceration, a technique commonly employed in other anatomical settings (Todd, 1923; Hunt and Albanese, 2005; Dayal *et al.*, 2009).

Skeletisation by *Dermestes maculatus* hold some other prime advantages: skeletons of known origin are retained with consent for teaching and research purposes; the skeletons can be readily replaced as necessary; and the newly produced specimens have much higher quality

Table 1.   Details of well-known osteology collections from across the globe.

Skeletal Collection	Founding Institution	Anatomy Department	Current Location	Processing Method	Skeletisation Method	Degreasing Method	Sample Size	Actively Adding New Samples
Hamann-Todd Human Osteological Collection[a]	Western Reserve University, Cleveland, OH, USA	YES	Cleveland Museum of Natural History (OH, USA)	Indoors	Darsis & Steam Maceration	Trichloroethane	3,714	NO
Robert J. Terry Anatomical Collection[b]	Missouri Medical College, Washington University, MO, USA	YES	Smithsonian National Museum of Natural History (DC, USA)	Indoors	Darsis & Hot Water Maceration	Benzene Vapors	1,728	NO
WM Bass Donated Skeletal Collection	The University of Tennessee, Knoxville, TN, USA	NO - Forensic Anthropology Center	The University of Tennessee, Knoxville, TN, USA	Outdoors	Surface Exposure and Cold Water	Occasional Use of Heated Water	1,372	YES
Pretoria Bone Collection	The University of Pretoria, Pretoria, South Africa	YES	The University of Pretoria, South Africa	Indoors	Darsis & Hot Water Maceration	Trichloroethylene	1,433	YES

	The University of Witwatersrand, Johannesburg, South Africa	YES	The University of Witwatersrand, Johannesburg, South Africa	Indoors	Darsis & Hot Water Maceration	Trichloroethylene	2,605	YES
Raymond A. Dart Collection of Human Skeletons[c]								
Chiba Bone Collection	Chiba University, Chiba, Japan	YES	**No longer accessible	Outdoors	Darsis & Surface Exposure	Sun Bleaching	205	NO
Khon Kaen University Osteological Collection	Khon Kaen University, Khon Kaen, Thailand	YES	Khon Kaen University, Thailand	Outdoors	Darsis & Sand Burial	Sun Bleaching	>700	YES
Chiang Mai University Skeletal Collection[d]	Chiang Mai University, Chiang Mai, Thailand	YES	Chiang Mai University, Thailand	Indoors	Darsis & Sand Burial	Sand Burial	>1,000	YES

[a]Information from Kern (2006).
[b]Information from Hunt and Albanese (2005).
[c]Information from Dayal et al. (2009).
[d]Information from King et al. (1998).

because: (1) the skeletisation can be tightly controlled; (2) elements can be systematically tracked without commingling; and (3) older individuals tend to participate in university body donor programs so truly adult morphology can be saliently demonstrated to students — for examples see Stephan and colleagues (2016). By broadening body donor acceptance criteria (i.e., lifting the institutional bodyweight restriction by circumventing fluid-weight addition during embalming), we are able to better meet the intent and wishes for a broader range of school registered body donors.

TOWARDS THE KOLKATA MEMORIAL ASSEMBLAGE

The generation of complete non-comingled skeletons from indefinitely bequeathed bodies of body donors offers the retirement of unknown-provenance skeletons from day-to-day undergraduate teaching programs and allows it to be housed instead in a memorial assemblage. This assemblage will hold the potential for future repatriation should forensic techniques ever enable it, and the assemblage will serve in the meantime as a recent and tangible reminder to students (and staff) of the importance of consent and responsibilities anatomists hold towards their fellow humans.

The removal of the unknown-provenance material from day-to-day undergraduate teaching will confer a heightened level of respect to the concerned individuals, paying them memorial tribute. The memorial assemblage may also house other South-Asian osteology returned to the University from the general public, as former medical practitioners retire.

The indefinite body donor skeletization program at UQ was initiated in late-December 2014 and at the end of July 2016 ten skeletons have been completed (Stephan *et al.*, 2016). Over the next five years we anticipate 35 skeletons to be produced, enabling their teaching use in single classes as large as 210 students. Ongoing skeletisation will open up possibilities to generate a research sample, and replacing teaching samples with new entries as required into the future. Although some deterioration of the bones is inevitable with daily handling, steps are taken to reduce the impact of handling by foam padding in storage boxes, separate storage of skulls and os coxae from the rest of the infra-cranial skeleton (Fig. 14), and staff (not students) packing and storing skeletons after classes.

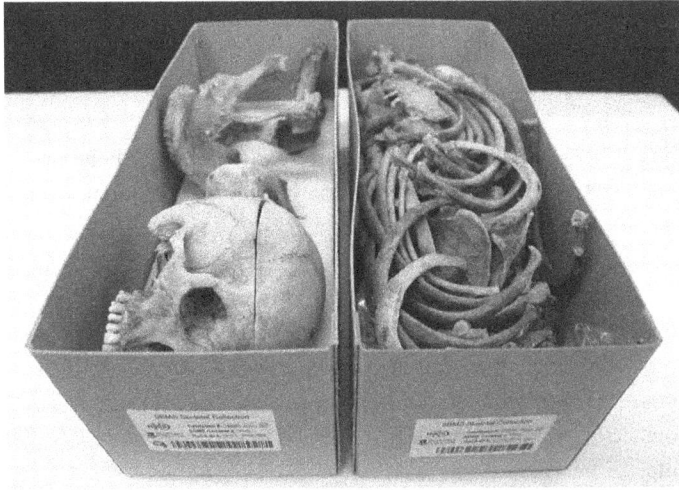

Figure 14. Example of indefinite body donor skeleton storage on foam in two boxes.

Consequently, over the same 5-year time frame, it will be possible to simultaneously retire the entire unknown-provenance material from day-to-day undergraduate teaching, a process that has already begun with inventory of the unknown-provenance remains and production of new skeletons from indefinite body donors.

Eliminating commingling of the unknown-provenance individuals, as much as current forensic anthropology techniques permit, is paramount in ensuring we have a full appreciation of the material contained therein and to enable tracking into the future. It is our intent that the memorial assemblage will not be shelved or hidden out of sight but rather, stored in a respectful and transparent way, providing a daily reminder to both students and staff of past (but not so distant) events and reaffirming the ever pressing requirement for human respect and dignity in our scientific endeavors.

REFERENCES

AFP. 1985. Police seize skeletons in child murder inquiry. *The Age*, 25 July 25.
Anadrabi J. 2009. Ban fails to stop sales of human bones. *The National*, 13 February.

Anonymous. 1829. The late horrible murders in Edinburgh, to obtain subjects for dissection. The *Lancet* 11:424–431.

Anonymous. 1976. We've all got one but there is now a shortage. *New Scientist*, 8 July.

Anonymous. 1981. Sell Skeletons. *The Montreal Gazette*, 21 February.

Anonymous. 1985a. Human skeleton trade blacklisted by India. *The Montreal Gazette*, 17 August.

Anonymous. 1985b. Mistaken mob burns to death tourist in India. *Orlando Sentinel*. United Press International. 7 September.

Anonymous. 1985c. Mob kills Australian tourist in India. *UPI*, 6 September.

Anonymous. 1989. Calcutta exports skeletons. *Manila Standard*, 17 April.

Anonymous. 2007a. Bone trader in biz for decades. *The Times of India*. 25 April.

Anonymous. 2007b. Skeleton 'trade' turns bone of contention. *The Telegraph*, Calcutta, 27 February.

Anonymous. 2014. Bone factories are back in business. *The Day After*, India. 1–15 April.

Ball JM. 1928. *The Sack-'Em-Up Men: An Account of the Rise and Fall of the Modern Resurrectionists*. Edinburgh: Oliver and Boyd.

Banerjie I. 1985. Skeleton exports: bizarre trade. *India Today*, 30 November.

Carney S. 2011. *The Red Market: On the Trail of the Wolds Organ Brokers, Bone Thieves, Blood Farmers, and Child Traffickers*. New York: HarperCollins.

Chatterjee A, Sarkar I. 2007. Unbreakable bone baron. *The Telegraph*, Calcutta, 24 April.

Chaturvedi S. 2015. More than 100 dead bodies found in Ganges. *The Wall Street Journal*, 14 January.

Cole H. 1964. *Things for the Surgeon: A History of the Extraordinary Era When Body-Snatching Gangs Carried on a Grisly Trade With the Most Eminent Surgeons in the Country*. London: Heinemann.

Corrales S. 1987. Skeleton shortage is giving medical schools a headache. *Los Angeles Times*, 21 April.

Dayal MR, Kegley ADT, Strkalj G, Bidmos MA, Kuykendall KL. 2009. The History and Composition of the Raymond A. Dart Collection of Human Skeletons at the University of the Witwatersrand, Johannesburg, South Africa. *Am J Phys Anthropol* 140:324–335.

Fineman M. 1985. India: A serene, spiritual mecca has become a nation of assassins. *Chicago Tribune*, 27 September.

Fineman M. 1991a. Changing lifestyles: Living off the dead is a dying trade in Calcutta. *Los Angeles Times*, 19 February.

Fineman M. 1991b. Outrage brings halt to Calcutta's human-skeleton trade. *Los Angeles Times*, 2 June.

Fisher W. 1943. The Bengal Famine: 50,000 Indians weely succumb to disease and starvation in spreading catastrophe. *Life Magazine*, 22 November.

Franklin KJ. 1949. *A Short History of Physiology*. London: Staples Press.

French R. 1997. The anatomical tradition. In: Bynum WF, Porter R (eds) *Companion Encyclopedia of the History of Medicine*. London: Routledge, pp. 81–101.

Gill G, Burrell S, Brown J. 2001. Fear and frustration — the Liverpool cholera riots of 1832. *The Lancet* 358:233–237.

Gray H. 1858. *Anatomy Descriptive and Surgical*. John W. Parker and Son.

Gupta J. 2011. Seized skeleton may be the last one in market. *The Times of India*, 18 February.

Guttmacher AF. 1935. Bootlegging bodies: A history of body-snatching. *Bull Soc Med Hist Chic* 4:353–402.

Hefner JT, Spatola BF, Passalacqua NV, Gocha T. 2016. Beyond taphonomy: exploring craniometric variation among anatomical material. *J Forensic Sci.* 61: 1440-1449.

Himmerlmann L. 2007. From barber to surgeon–the process of professionalization. *Sven Med Tidskr* 11:69–87.

Hunt DR, Albanese J. 2005. History and demographic composition of the Robert J. Terry anatomical collection. *Am J Phys Anthropol* 127:406–417.

Hunter RH. 1931. *A Short History of Anatomy*. London: John Bale, Sons and Danielsson.

Hurren ET. 2014. *Dying for Victorian Medicine: English Anatomy and its Trade in the Dead Poor, c. 1834–1929*. New York: Palgrave MacMillan.

Kern KF. 2006. T. Wingate Todd: pioneer of modern American physical anthropology. *Kirtlandia* 55:1–42.

King CA, Iscan MY, Loth SR. 1998. Metric and comparative analysis of sexual dimorphism in the Thai femur. *J Forensic* Sci 43:954–958.

Litten W. 1944. Bones for sale. *Life Magazine*, 21 February.

Lockhart RD. 1927. The art of learning anatomy. *The Lancet* 210:460–461.

MacDonald H. 2005. *Human Remains: Episodes in Human Dissection*. Melbourne: Melbourne University Press.

MacDonald H. 2006. *Human Remains: Dissection and its Histories*. New Haven: Yale University Press.

MacDonald H. 2010. *Possessing the Dead*. Melbourne: Melbourne University Press.

Major RH. 1954. *A History of Medicine*. Springfield: Charles C Thomas.

Majumdar B. 2006. India cops uncover human 'bone factory'. *Independent Online*, 6 March.

McBride P. 2014. The Pyres of Varanasi: Breaking the Cycle of Death and Rebirth. *National Geographic*, Proof, 14 August.

McConnaughey J. 1986. Skeleton supply down as India stops bone sales. *The Sumter Daily Item*, 26 Feb.

McMaster University of Health Sciences. 2009. So...*just where do those skeletons come from anyway?* MacAnatomy: Education Program in Anatomy.

Moore W. 2005. *The Knife Man: Blood, Body-Snatching and the Birth of Modern Surgery*. London: Bantam Press.

Payne JA. 1965. A summer carrion study of the baby pig *Sus Scrofa Linnaeus*. *Ecology* 46:592–602.

Putka G. 1986. Indian Export ban leaves medical schools bone-dry. *Wall Street Journal* via Montreal Gazette, 12 June.

Queensland Government. 2016. Transplantation and Anatomy Act 1979 (Current as at 24 March 2016). Brisbane: Queensland Government.

Richardson R. 2000. *Death, Dissection and the Destitute*. Chicago: The University of Chicago Press.

Robinson JO. 1984. The barber-surgeons of London. *Arch Surg* 119: 1171–1175.

Ross I, Ross CU. 1979. Body snatching in nineteenth century Britain: from exhumation to murder. *British Journal of Law and Society* 6:108–118.

Russell KF. 1973. Anatomy and the Barber-Surgeons. *Med J Aust* 1: 1109–1115.

Russell M. 2011. Skeletal story behind these lovely bones. *The Age*, Melbourne, 29 May.

Scheuer L, Black S. 2000. *Developmental Juvenile Osteology*. London: Academic Press.

Shapiro SL. 1972. The skeptical anatomist. *Eye Ear Nose Throat Mon* 51: 103–107.

Shultz SM. 1992. *Body Snatching: The Robbing of Graves for the Education of Physicians in Early Nineteenth Century America*. Jefferson: McFarland & Company.

Singer C. 1957. *A Short History of Anatomy from the Greeks to Harvey*. New York: Dover.

Singer C, Underwood EA. 1962. *A Short History of Medicine*. Oxford: Oxford University Press.

Spink K. 1992. *The City of Joy*. London: Random House.

Stephan CN, Caple J, Veprek A, Sievwright E, Kippers V, Moss S, Fisk W. 2016. Whole Body Skeletisation of Indefinite Body Donors at UQ using *Dermestes Maculatus*. In preparation.

The Select Committee on Anatomy. 1828. *Report Ordered by The House of Commons*. Great Britain: Parliament of Great Britain.

Todd TW. 1923. The effect of maceration and drying upon the linear dimensions of the green skull. *J Anat* 57:336–356.

Tward AD, Patterson HA. 2002. From grave robbing to gifting: cadaver supply in the United States. *JAMA* 287:1183.

6

THE CURRICULUM OF COMPASSION: CONNECTING STUDENTS WITH DONOR FAMILIES

Ernest F. Talarico, Jr.

Department of Anatomy & Cell Biology, Indiana University School of Medicine — Northwest Campus, Gary, Indiana 46408-1197, USA
etalaric@iun.edu

ABSTRACT

Physician training typically begins with dissection of anatomical donors, which is commonly used to teach structure and function. However, cadaveric dissection also provides opportunities for multifaceted educational experiences that build other skills and competencies critical to patient care. This chapter discusses a teaching paradigm in which students interact with families of the deceased persons whom they are dissecting. This novel approach represents a shift from the traditional paradigm, which emphasizes students' clinical detachment, and restores identity to anatomical donors by treating them as "first patients." This practice establishes the patient as the primary focus of medicine early in the curriculum. From patients' arrival in the laboratory, through donor memorial services and even after the return of cremains, the "student-patient-family relationship" evolves, building connections and developing skills that directly affect future patient care. Learning anatomy and medicine through the first patient is achieved according to five guiding principles: First Patient; Knowledge; Reflection and Reflective Practice; Treating the Total Patient; and Professionalism. Here, these principles and their implementation are described, and their impact on student doctors and learning outcomes are discussed. Results suggest that mastery of basic science knowledge and competencies, including professionalism, compassion, and leadership skills, are enhanced by this protocol.

INTRODUCTION

Today, medical school curricula are undergoing dynamic revisions that are decreasing didactic contact time and increasing time for self-directed learning, physician shadowing and experiences designed to enhance competencies. Yet, the material to be learned is voluminous and the pace rapid. Human gross anatomy — an essential and foundational medical school course, and one most often approached by students with uncertainty and apprehension — is undergoing one of the most dramatic transformations, a culture change or paradigm shift that might be termed "humanistic anatomy" (Dyer and Thorndike, 2000; Talarico, 2013; Štrkalj, 2014; Reilly, 2015). Gross anatomy is often the student doctor's first class subject and often represents his or her initial confrontation with death, dying, and end-of-life issues. The majority of student doctors have never dissected a human being prior to medical school. There, most anatomical gift programs provide students with little or no information about the deceased persons whom they dissect — giving cadaver donors numbers, rather than names, and an assigned cause of death (Terry, 2014). This practice is consistent with the traditional view that anonymity lessens the emotional stress of the anatomy laboratory experience and furnishes students with distance and objectivity as they learn basic sciences. Critically, this approach also removes the "patient" as the focus and relegates the donor to the impersonal status of "cadaver specimen" or "cadaveric materials."

Physicians have always needed mastery of basic science knowledge as well as competencies such as professionalism, compassion, leadership and teamwork skills. These attributes are even more critical in the current medical industry, in which resilient learning institutions and lifelong learners must adapt with ever-increasing speed to shifting business, regulatory, and competitive environments. Additionally, the delivery of health care has become a modern group practice rather than a single-provider responsibility; this model requires a physician to be a team member and often the leader of several teams that must work together.

In the face of this dynamic landscape, medical school gross anatomy courses provide an environment for multifaceted educational experiences where dissection of the human body is used to teach structure and function as well as the psychosocial issues critical to developing the whole physician (Terry, 2014). Thus, the gross anatomy laboratory can be a unique environment that offers new approaches early in medical education that foster learning, leadership, and ongoing reflection on medical practices, all of which are instrumental in improving the quality of patient care. Further, the increasingly popular practice of hosting memorial ceremonies in which students thank anatomical donors and their families allows for reflection and represents a major shift in the training paradigm in medical schools with respect to the attitude and composition of physicians (Jones *et al.*, 2014). At Indiana University School of Medicine — Northwest (IUSM-NW; Gary, Indiana U.S.A.), a relatively recent approach upends convention by involving the families of those being dissected in the anatomy laboratory with the students who are performing the dissection (Talarico, 2013). The rationale behind this protocol is that information from families can augment medical knowledge and promote research, and that exposing medical students to grieving relatives can enhance the skills and competencies of these future physicians. This novel paradigm creates an interdisciplinary learning environment focused on the "donor as patient, teacher, and human being."

A NOVEL PROTOCOL

Typically, anatomical disciplines have preached a philosophy of detached concern and anonymity. In contrast, this novel approach shifts from the traditional paradigm and gives back "identity" to cadavers (i.e., donors), thereby returning the patient to the focus of medicine early in the curriculum. Further, this practice goes far beyond the incorporation of a memorial service but utilizes an all-encompassing teaching protocol in which *the teaching philosophy itself is a commemoration and tribute to the donor*. The implementation of this protocol begins prior to the arrival of donors on campus with discussions that focus on the history of cadavers

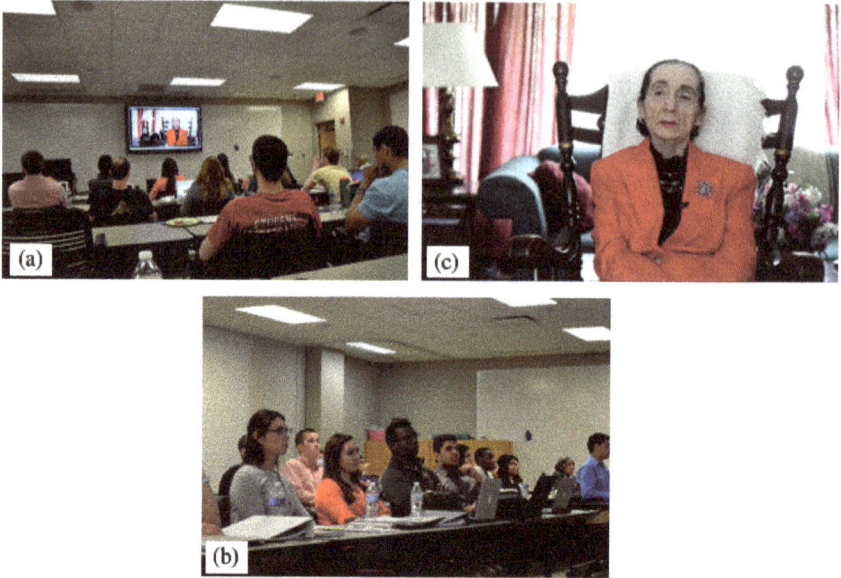

Figure 1. Listening to, and watching, messages from First Patients. (a) and (b) Medical students listen to recorded messages left for them by first patients whom they will be dissecting in the human gross anatomy laboratory. Following each message there is discussion between students, peers and faculty about the message and the thoughts of students. (c) Patricia A. Kelly tells students why she and her husband donated their bodies for their use in medical education, and she explains how important the letters are that students write to families in helping the families and treating the total patient.

in anatomical education and medicine, as well as on the anatomical donation process. Students listen to videos of donors (Fig. 1) explaining why they desire to bequeath their remains for anatomical education, and to videos of families who have been involved with prior students during dissection. Students become aware that human dissection is a fascinating learning experience in which they are privileged to participate, and they begin to recognize their role(s) as doctors in treating patients (and patients' families). Tasks such as transport, storage, and maintenance of donors are regarded as professional obligations and responsibilities to the donors and donors' families. Student doctors are confronted with the realization that this patient is a person whose full name they now know and with whom they will become intimately familiar. Critically, this approach also allows contact between survivors of the donor's family and the student doctor who is dissecting the donor from the onset of, and throughout, the formal course in human gross anatomy. Students learn more about their patient — both medical history and lifestyle — and they learn more about compassion and professionalism in total patient care. Still further, contact with families of the deceased illuminates other conditions or topics that lend themselves to further research for lifelong learners. Data shows that this unique approach not only results in an increase in student scores on the *National Board of Medical Examiners*® Gross Anatomy and Embryology Subject Examinations, but also creates relationships between students and first patients' families that positively impact the health care provided to future patients (Talarico, 2013). The five guiding principles of this educational method that enhance physician competency are discussed herein (Table 1).

Table 1.   Guiding principles of the novel paradigm.

PRINCIPLE	DESCRIPTION
"The First Patient"	The human cadaver will not be viewed as a specimen, but as my first patient, my donor and a human being.
"Knowledge"	With my first patient as my focus, I will do my best to learn all that can be learned.
"Reflection and Reflective Practice"	Through contact with my first patient and the family members of my first patient, I will reflect upon my patient's life and the gift(s) he/she has given to me so that I will be a better physician.
"Treating the Total Patient"	I will use all that I learn from my first patient to understand the medical and non-medical factors that contribute to a patient's total well-being and quality of life.
"Professionalism"	With my first patient as my focus, I will learn professionalism, respect for human life and human dignity, and integrate these qualities that we expect of our best physicians beyond the anatomy laboratory.

GUIDING PRINCIPLES

"The First Patient." In the present protocol, the anatomy laboratory is considered as the student's first clinical experience, with the human cadaver donor as the student doctors' first patient and best teacher (Shapiro and Talbot, 1991). It is developed to the furthest extent through the aspects of *responsibility, name, report, gift, and communication.* Students at IUSM-NW are the only custodians of their first patients (Talarico and Walker, 2007; Talarico, 2010), and each actively receives his/her patient, and assumes responsibility for transport, storage, maintenance, and preparation of remains for return to families. These tasks are regarded as professional obligations (or responsibilities) to the patients and patients' families. These responsibilities are comparable to those of a practicing physician, who has a duty to care for his/her patient in the best way possible.

In the New Zealand-based film *Donated to Science* (Trotman and Nicholson, 2009) and in the Trinity College (Dublin, Ireland) documentary, *A Parting Gift* (Devereux and Foley, 2014), students refer to the cadaver as "it," and donors are not introduced by name on the first day of anatomy laboratory — "We were standing around looking at 'it'." Another student said that the body "used to be a person. Now you're going to cut them up." Furthermore, students featured in these films do not meet donor family members until long after the formal course in gross anatomy has concluded. Here, we suggest that awareness and use of the patients' full names from the beginning of the anatomy experience maintains humanity in the learning process and keeps the focus of study and medicine on the patient (Table 2).

The third aspect of "the first patient" principle is an "autopsy report". Students perform full-body examinations on their patients, as part of an "autopsy report" that they develop throughout the course of dissection. This report includes diagrams and narrative boxes to record patient data for every region and organ of study, as well as a "radiology" section that requires students to interpret and report on normal and abnormal findings from films obtained via hands-on, whole-body medical imaging (i.e., X-ray, ultrasound, CT Scan, and MRI Scan) performed with their student counterparts in Radiography. Laboratory observation is augmented with the application of knowledge and information gained from family contact

Table 2.   Reflective Writings. These reflective writings from student doctors and physicians document positive outcomes of the principles outlined in this educational protocol.

"Knowing Edwin and his family has exposed me to a professional way of dealing with a patient and his or her family. For me personally, this has been a life-changing experience. I could not possibly express how grateful I am for the communication and interaction with Edwin and his family members. It has opened my eyes in many different ways and exposed me to the true definition of patient care. You and Edwin have made a drastic impact on me and other medical students. The anatomical knowledge and skills for compassion, empathy and communication will help me to be the best possible physician."
(First-Year Medical Student)

"As a medical student, I am given a curriculum which will provide me with the knowledge to treat an ailing patient. Although thorough knowledge of a disease is vital in providing a cure, knowledge of the patient is equally necessary. Moreover, to know and understand a patient, a relationship must be established. This is why a curriculum of compassion is extremely critical. In my experience, knowing my donor's name helps to establish this relationship with both patient and patient's family. I had the privilege to interact with the donor family of my first patient and I was able to learn about the history of my patient. I learned how much my donor meant to their family and how important it was that I treated my donor in a way that was specific to the patient's and the family's expectations. Furthermore, the time I spent with the family was comforting and it facilitated solace, even contentment. I am grateful for participating in the student-patient-family relationship because it has taught me the importance of kindness, sympathy, and understanding. As a result of my experience, I believe I will become a better doctor because it has guided me in treating a patient, not just a disease."
(Second-Year Medical Student)

"During first-year anatomy laboratory, getting to know my patient and her family turned out to be such a blessing to my life. Initially, I thought it would be a bit unnerving to get to know the family of the person who donated their remains for my education. But to my surprise, my patient's family was not only excited to get connected to the students who would have this interesting bond with their loved one, but willing to embrace the students involved with their loved one as extended family. I also grew to have reciprocal feelings. And these feelings continued on although the anatomy course came to an end. I continued to fellowship with the family over great home-cooked dinners set up in their dining room. Each time that I step into an examination room and for each clerkship rotation, I remember this. One might consider having such a personal relationship with the family of an anatomical donor as odd or inappropriate; however, I believe that fostering such relationships helps medical students keep the value of life and relationships as an important perspective in the treatment and care of patients."
(Fourth-Year Medical Student)

(*Continued*)

Table 2.   *(Continued)*

"I had the opportunity to meet Dorothy's husband and son. While I understood and respected the life behind this 'study aid', it was then that I actually felt the soul, the vitality, behind Dorothy. And when Dorothy's husband showed us a stunning picture of her younger self looking into the distance and another one of her, smiling from a year ago, I could not help seeing more than life behind those eyes. Dorothy was now a woman who had lived. This cadaver...this woman... had left behind 11 children and a loving husband. She had lived through a period that you can either read about through history books or hear from talks with grandparents. Our lives are connected in a way, Dorothy's and mine. I learned lessons far beyond the laboratory and textbooks that I now put into practice with each patient and patient family that I treat."
(Resident Physician)

(described below), and this information (like a history taken by a physician) is added to this report using anatomical and medical terminology. Collectively, this stimulates discussion at the patient (i.e., cadaver) -side of anatomy and ethical topics (i.e., death and dying, end-of-life issues, illness/death and the family, etc.), as well as the teaching of critical concepts that transcend academic learning, such as professionalism, respect, and human dignity.

The fourth aspect of "first patient" is the patient as "gift." In the documentaries *Donated to Science* (Trotman and Nicholson, 2009) and *A Parting Gift* (Devereux and Foley, 2014), students express fears that dissections will make them emotionally cold people, and another reflects that, "One thing that I realized was how easy it was to start seeing the body as a tool of learning rather than as a person. And so, after dissecting, I have to remind myself again of how this was a live person." The protocol at IUSM-NW addresses this concern by emphasizing the donor's humanity from the start through "practice" with the cadaver as "first patient" and "human being." The "practice" of "first patient and human being" has further implications for the developing physician. It is an important step in the professionalization of medical students and incorporates transformative aspects in the laboratory and at the patient's side that are rarely considered in medical curricula. By caring for their patients and learning appreciation for the gift of self, students learn attributes that they will need as physicians: compassion in dealing with death; knowledge about end-of-life issues and the limits of medicine; and respect for life and human dignity. Further, they learn about themselves, as they face the realities of life and death and the limitations of their chosen profession. It's their first step to becoming a doctor. Here, we emphasize that the cadaver as first patient is a gift, and that the act of receiving this gift commits one to a relationship — the doctor–patient relationship. Though medical technology and clinical expertise are pivotal to contemporary medicine, the doctor–patient relationship and successful face-to-face relief of actual suffering continues to require the application of ordinary human feeling and compassion, known as bedside manner (Flocke *et al.*, 1998).

The fifth aspect is "communication and exchange of information with the patient's family." An important part of this paradigm is communication with family and friends of the first patient. This is discussed in more detail under "Treating the Total Patient"; it suffices here to state that this is not unlike communicating with a living patient's family in the examination room.

"Knowledge." With acceptance of the gift of a first patient comes responsibility, and what one does, and intends to do, with one's first patient and the knowledge gained from this patient, constitutes the basis for that responsibility. One way students can demonstrate respect for this patient is by working as hard as they can and learning as much as possible from their ultimate teacher, the cadaver donor (Talarico and Prather, 2007; Talarico, 2010). Student doctors have a three-fold obligation: (1) to make the personal commitment to take fullest advantage of the gift that has been placed before them; (2) to learn as much as possible about the human body from their first patient; and (3) to perform the dissection to fulfill the mandate to become as knowledgeable and competent as one is able (Sukol, 1995). Thus, performing dissection or anatomical research that extends outside of the anatomy laboratory, in the pursuit of knowledge, is a direct extension of the student's commitment to becoming a competent physician, and is in itself commemoration and memorialization of the first patient.

In the present protocol, the inclusion of an "autopsy" has been previously discussed. Although the inclusion of hands-on radiography might be a separate chapter (Gunderman and Wilson, 2005), it is worth discussing here in more detail because of the degree to which our students are involved. Not only do student doctors work in interdisciplinary format with their student radiographer counterparts in obtaining high-quality images of their patients, but they also learn how to use computer software to manipulate and interpret images, as well as create other constructs from multi-planar or three-dimensional reformatting. Radiographic series from their patients are available to students during dissection, facilitating the integration of gross anatomy and radiology via direct application.

Dissection, hands-on radiography, and the inclusion of clinical application activities for each region dissected (i.e., total knee arthroplasty — lower extremity; endotracheal intubation — thorax; blood draws — upper extremity; ultrasound — abdomen and pelvis; Talarico, 2010) enable students to commemorate the donor by using the gift to the fullest extent. Information gained from radiographs (not otherwise available) or from direct communication with the patient's family, in addition to variations, pathologies, or anomalies that are discovered, often give rise to additional opportunities for continued learning. In my own laboratory, examples of enhanced student learning that have resulted from discoveries in first patients include cadaver-based research that focuses on hyperostosis, malignant vaginal mucosal melanoma, renal cell carcinoma, dysexecutive syndrome, ineincephaly, diabetes, Lynch syndrome, Dandy–Walker malformation, intracerebral hemorrhage, kyphoscoliosis, giant hiatal hernia, and medical humanities (Talarico and Prather, 2007; Talarico *et al.*, 2008; Talarico and Buchler, 2011; Talarico and Koveck, 2012; Talarico and Hiemstra, 2013; Talarico and Frantz, 2013; Talarico and Platt, 2014; Talarico and Vlahu 2015).

One final note: knowledge is community. Obligation to community involves the student's recognition of having received an education in large part from one's first patient, and the affirmation of the first patient and community, such that students recognize that it is society (i.e., community) that allows for anatomical donation. Thus, students are encouraged to reciprocate this gift of knowledge to the community via competent patient care, by being an educator or mentor, and through service-learning activities.

"Reflection and Reflective Practice." The foundation of confidence and belief in one's self has its basis in reflection (Forrest, 2008). Reflective Practice is the capacity to reflect on action engaging in a process of continuous learning, which is one defining characteristic of professional practice (Schon, 1983). Thus, physicians and student doctors are reflective practitioners or those who, at regular intervals, look back at the work they do, and at the work process, and consider how they can improve. Reflective practice is an essential tool in clinical settings where physicians learn from their own professional experiences, rather than from formal teaching or knowledge transfer. In fact, reflective practice may be the most important source of professional development and improvement following graduation and residency. Thus, it can be suggested that in order to succeed as professionals, medical students must engage in reflection and reflective practice, and this begins in the anatomy laboratory.

The importance of reflecting on what you are doing, as part of the learning process, has been emphasized by many investigators and is part of the adult learning cycle (Kolb, 1984). This model suggests that there are four stages in learning which follow from each other (Fig. 2): (1) Concrete Experience that contributes to student motivation to learn;

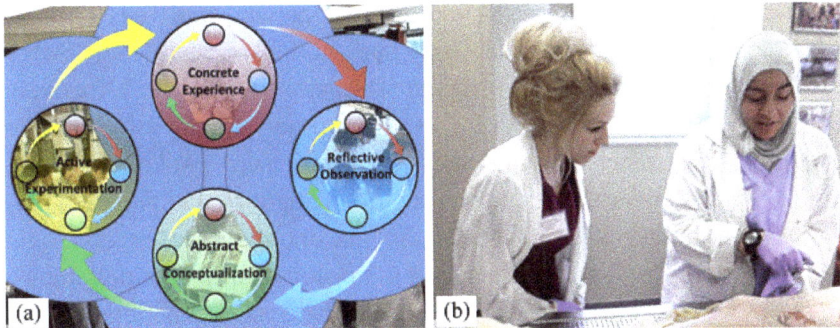

Figure 2. Learning model for reflection and reflective practice. (a) This model suggests a four-stage, cyclic learning process. Concrete experience is followed by reflection on that experience on a personal basis. This is followed by abstract conceptualization, or the derivation of general rules describing the experience, and may involve the application of known theories to the experience. Now, the learner can formulate methods for altering the next occurrence of the experience, known as active experimentation, leading to the next concrete experience. Depending on the topic, there may be cycles within cycles occurring at the same time. (b) A medical student team leader (left) instructs another student about the lower extremity, and discusses this relative to her grandmother's lower extremity problems, who by chance happened to be an anatomical donor at this same table three years prior.

(2) Reflection of that experience; (3) Abstract Conceptualization, in which there is derivation of general rules describing the experience; and (4) Active Experimentation, the construction of ways of modifying the next occurrence of the experience. The current paradigm, actively encourages this sort of reflective learning in the anatomy laboratory via discussion during dissection of the first patient and through reflective writings (Talarico, 2013; Wagoner and Romero-O'Connell, 2009) and letters to families of first patients. The implementation of this principle allows for questioning and planning for improvement to begin early in the educational process. Typical questions might be: How can I improve my skill for dissection? How can I more effectively communicate and work with my team members to accomplish goals? How can I more effectively communicate with my patient's family (and my patient) to understand my patient's condition and all factors in my patient's health and well-being?

"Treating the Total Patient." Total patient care is a cooperative multidisciplinary team approach to treating each patient's specific needs, allowing the patient to reach a state of optimum health, well-being, and productivity in society. Implementing total patient care (i.e., patient-centered care) requires a fundamental shift in thinking from how best to provide a wide variety of independent services to how to effectively combine individual service components into an integrated health care experience that meets patient needs and preferences (Wakefield *et al.*, 1994). This involves the understanding of a patient's "traditional" medical issues, but also psychosocial, socioeconomic, and occupational issues, as well as familial history, relationships, and support systems. In a sense, the physician also treats the family as part of the total patient care process.

The concept of total patient care can be applied in the anatomy laboratory, where the first patient is a nonliving patient. Because the patient is the *common denominator of medicine* (Engel, 1977; Aziz *et al.*, 2002), it is suggested here that an understanding of total patient care can be enhanced by interaction/communication with the first patient's family. This contact does not occur in the laboratory during dissection but comes via written letters, telephone conversations, and email, and then in the laboratory during the annual memorial service.

Early in the anatomy course, students write letters to the families of their patients; this is the beginning of ongoing correspondence that constructs the stories of past lives, humanizes the "cadaver experience", and enables student doctors to learn more about the lives of those who gave the ultimate gift of self to help others. In these letters, students tell families about themselves and explain their pledge to their first patient to learn all that can be learned and to treat their loved one with the utmost respect and dignity. Students learn what questions to ask family members and how to ask them, not only gaining anatomical and clinical information about their first patient, but also applying this information via reflective practice to increase understanding of basic science and clinical concepts. Survivors of the deceased also ask questions about the student doctor or about findings in their loved ones and how these findings might affect their own health. Students are extensively counseled by faculty regarding their professional and ethical obligations, and letters are reviewed and discussed with students prior to contact with family members.

Typically, families forward medical records, narratives, photographs, and videos, thus enabling students to learn about the medical history, life, and family relationships of their patients, and allowing students to share in personal moments such as birthdays, weddings, child births, etc. All information is shared in class and letters are read aloud. Secondary history about past illnesses, surgeries, conditions, and causes of death enables students to delve deeper into laboratory findings and interdisciplinary care. Social, occupational, and lifestyle information allows students to appreciate the influence of multiple factors throughout the life of their patients that impacted their total health and well-being. Often, this information can be directly applied in the laboratory to various conditions: smoking and lung disease; poor diet and/or lack of exercise with obesity/wasting and/or heart disease; hyperlipidemia and/or atherosclerosis and pulmonary embolism or stroke; arthritis and/or fracture and prosthetic implants; and environmental/ occupational exposures and cancers. Ultimately, these relationships climax in an annual Service of Thanksgiving & Remembrance of Our Donors, organized and conducted by students. For some students, contact with family members continues through clinical training and beyond (Talarico and Prather, 2007; Talarico, 2010; Lewis-West and Hiatt, 2011; Talarico, 2011).

Support for this method of enhancing students' understanding of total patient care comes from several sources. First, Flock *et al.* (1998) examined family as a component of patient care, and found that 72% of primary care patients report that multiple family members see the same doctor. Further, in one-third of observed visits, a patient's family member was present, and in 18% of visits, physicians discussed health problems with a patient's family member. Thus, through our learning approach involving the families of first patients, students gain valuable insight and practice into treating the total patient via interaction with the first patient's family. This also constitutes a basic foundation of understanding the social and community contexts of health care, but is initially centered on the cadaver as first patient. These sorts of interactions incorporate reflective practices and foster an understanding of the doctor–patient relationship and the importance of family/friends in medical care and total health (Fig. 3). Further, feedback from student doctors and families serves to document that, from this method, students come to understand the emotions and feelings of first patients and families; students therefore appreciate the importance of compassion in patient care, and learn about the social and spiritual components involved in patient health. Further, they report that additional benefits derived from this experience are taken into their future careers. Families also opine that this program has value to students and assists with closure both for the family and the students (Table 3).

Finally, *Donated to Science* (Trotman and Nicholson, 2009) and *Parting Gift* (Devereux and Foley, 2014) further illustrate the educational value of this protocol. In these documentaries, students repeatedly express their desire to gain more knowledge about their donors than what is typically made available to them. In addition, families state that they feel it is important for students to understand "who the donor was"; "what their life was," and that this "makes them appreciate the gift and makes them better doctors." The protocol described here yields a much better understanding of the first patient's life and personhood than what appears to be enjoyed by the majority of students at Otago and at Trinity College (Dublin, Ireland). In contrast, the students at Otago are not given their donors' first names at any time throughout their learning experience. At both institutions, students' only contact with families appears to take place at the memorial service, which means any value such connections might have brought to the learning process have been lost. At IUSM-NW, such

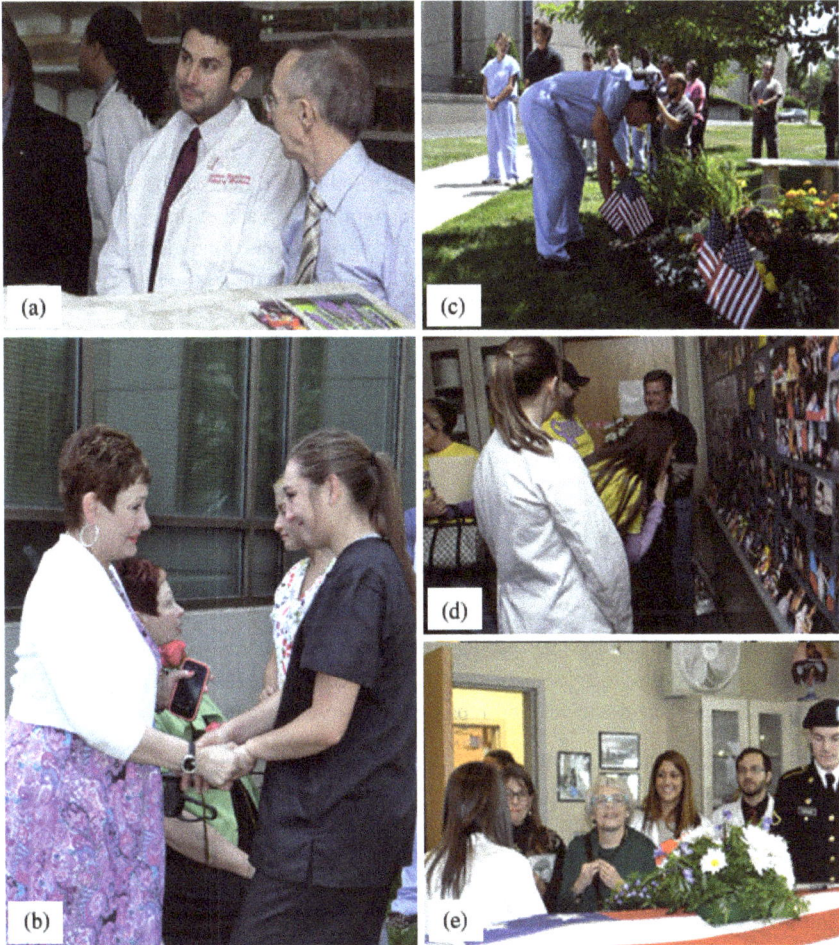

Figure 3. Interacting with, and learning from, the First Patient's Family. (a) With the son of his first patient, a student stands beside his patient's table and chats about his experience. (b) While comforting the daughters of their first patient, students talk about the patient's life and their goals as future physicians. Each thanked the daughters for their mother's gifts to them. (c) Student Doctors, faculty, physicians, and others place roses and flags "by name" for each first patient at an on-campus, donor memorial garden established by medical students. The grandson of a first patient (upper left) speaks about his grandmother and grand-father–both anatomical donors–as he begins this Service with student doctors. (d) Although they have been communicating with their patient's family for more than 5 months, students talk to the family about the patient's life and learn about numerous life events through a pictorial collage of their patient's life. (e) Five months after initially meeting her patient and family, a student doctor reads a letter of reflection and gratitude to the family of her first patient immediately preceding the presentation of military honors for her donor.

Table 3.   Thoughts from students and family members about this approach.

Comments from student doctors

"Through the student-patient-family relationship I was able to not only gain more clinical knowledge and engage in clinical research, but also connect on a personal level with my patients, Josh and Lydia, and their families. This helped me reflect on the affect [*sic*] my patients' lives had on their families, friends and communities, and of their gifts to myself. Now, their lives continue to affect people as they have motivated me to learn all that can be learned, to hold myself to a higher standard and become the best compassionate and knowledgeable physician I can be."

"I took a phenomenal amount of information from this program: clinical experience, how to treat donors, how to treat patients, how to work in a group setting. The program does a fantastic job of letting us know that the cadaver is just not a donor; they are a patient. It's really a Kool [*sic*] feeling that you are treating them just like you will treat your future patients."

"I think that it's really neat, the uniqueness that I can have a relationship on both the scientific level with my first patient and at the same time a personal level with the family. This really drives home the concept of humanizing medicine and science."

"I have been in contact with my patient's family on a weekly basis. This program has strengthened my desire to enter the field of family medicine, because this approach really helps you to understand how relationships are a priority in patient care."

"The interactions that I have had with my first patient's family are extremely vital to my development as a physician. From caring for my first patient, I have learned that there is more than a patient. There are family, friends; there is a lifestyle. This protocol has given me a unique insight into care of my patient down the road in my career."

Comments from family members and others

"We met the students who we had been talking with, and I will never forget it. I felt like they were all my family. They were so grateful for the donation of my husband's body. All programs all over the world should be like this."

"I read letters from students that my son was their first patient and how honored they were to be with him. Words just weren't enough to console my grief of losing my son. Until I met them all and saw first-hand what wonderful, caring, considerate future Doctors that they truly were. They had all of those assets and so much more! I felt all of those things and I had the honor of meeting them and knowing that those words rang true in each and every one of them."

(Continued)

Table 3. (*Continued*)

"I was so impressed by all the students who were so very interested in Josh's interests, jobs, activities, life style and each and every aspect of him. His family, his daughter, his siblings, like and dislikes. This I felt was a tremendous way to actually get to know their patient, without really knowing him at all. But again, connecting with their first patient in a way no other physician's [sic] normally do."

"Over two decades of reporting, precious few stories stand out like this program. The participants' respect and admiration for their "first patients" and their survivors is truly a celebration of life. I could not "cover" the story without being changed."

relationships are established during the gross anatomy course (Talarico, 2010; Lewis-West and Hiatt, 2011; Reilly, 2015).

"Professionalism." The art of medicine requires more than medical knowledge and procedural skill. The Medical School Objective Project (AAMC, 1998) states that the goal of medical education is to produce physicians who are prepared to serve the fundamental purposes of medicine. Thus, physicians must possess attributes that are necessary to meet their individual and collective responsibilities to society. Physicians must be altruistic, compassionate, truthful, knowledgeable, and skillful in communicating with and caring for patients (Hundert, 2001). Critically, teaching and dissection in the anatomy laboratory allow for the integration of professionalism (and role recognition) in early medical education (Rizzolo, 2002; Bryan *et al.*, 2005; Lachman and Pawlina, 2006; Pawlina *et al.*, 2006; Pawlina, 2006; Swartz, 2006; Louw *et al.*, 2009; Lewis-West and Hiatt, 2011). Escobar-Poni and Poni (2006) gathered descriptions of professionalism from leading medical organizations, grouping them into keywords: demonstrate, commitment, behaviors, and core attributes. Within this proposed program, these keywords are put into action during gross anatomy laboratory when students interact with first patients and the patients' families. Herein, the expectations for assessment are that:

1. The student doctor is required to **demonstrate** respect, compassion, and integrity for their first patient, as well as for the confidentiality of patient information. They should dissect with purpose and respect.

Discoveries about their patient uncovered during the course of dissection or revealed from contact with survivors of their patient should be treated as confidential. Finally, they should care for, and maintain, their first patient in the best possible condition.

2. The student doctor should understand that the human gross anatomy laboratory and the gift of one's own body for dissection constitute an important **commitment** on the part of society to the student of medicine. The student doctor's commitment to the first patient and his or her family is a commitment to be honest, to pursue scientific knowledge, to improve the quality of care, and to maintain confidentiality. Students of medicine act as professionals when they show a real commitment to treat the anatomy course as more than just an isolated basic science, but as a means for interdisciplinary learning toward improved total patient care.

3. The student doctor must actively master essential **behaviors and core attributes** that we expect of health care providers, such as: altruism; honor and integrity; respect; responsibility and accountability; excellence; scholarship and leadership; competence; commitment; ethics; self-regulation; and teamwork. Altruism is defined as a selfless concern for the welfare of patients, and focuses on a desire to help others without reward. Thus, it can be inferred that altruism as a component of a medical student's professional behavior develops in part through self-assessment (Bryan *et al.*, 2005; Boyle, 2006). Critically, working with the first patient in the gross anatomy laboratory and interacting with a first patient's family helps to develop characteristics of professionalism that he or she will use throughout his or her professional career. This is supported through work presented by Escobar-Poni and Poni (2006), which showed cadaver-based activities help to develop skills in commitment, responsibility, leadership, teamwork, integrity, communication, accountability, and excellence, among others.

Each expected outcome is assessed via formal competency evaluation and rated as satisfactory if the student is responsible, reliable, and honest; understands and respects professional roles and standards; interacts effectively with other health care professionals; and puts patient (and peer) needs ahead of personal needs. A student receives a rating of praise if the student stands out among peers/colleagues; is most responsible, reliable, and honest; careful and consistent in professional roles and interactions; leads and coordinates in interactions with other health care professionals; and is selfless in dealing with patients and others. Lastly, if a student is rated unsatisfactory, then he or she is given the opportunity to remediate this principle via activities focused on the deficient skill (i.e., reflective writing, HIPPA study, etc.).

DISCUSSION

This chapter has focused on a change in how first-year medical students are learning about the body in its most basic sense. Medical students are treating their donors not as anonymous cadavers, but rather as their very first patients. Critically, through the implementation of a five-part protocol, students learn not just anatomy and how their patient died, but they have an opportunity to learn how he or she lived by communications and meetings with the patient's family. Students also have the opportunity to conduct a medical history; complete detailed autopsy reports and radiological examinations; and ask questions of the patients' loved ones. In this way, they learn not only about anatomy and the signs and results of illness, but also about the roles of physician and family in patient care. Further, the students should be able to handle this approach and translate it into positive, total patient care, even in addition to the plethora of other information and responsibilities that student medical doctors face (Talarico, 2013). Recently, a fellow student was talking to Brittany Winn, a student from Kouts, Indiana, who has been interacting with first patients and their families through this approach for the last three years, at a memorial service commemorating first patients. She said, "'You are right. Everybody's going to be different after they leave this program.' This is the 'Talarico protocol': to teach future health care professionals the importance of teaching every patient — from the "first patient" on — with respect and dignity." (Giles, 2015).

ACKNOWLEDGEMENTS

The author most gratefully acknowledges the contributions and gifts of the donors and their families to medical education and scientific research, and the dedication and respect of the students toward their first patients and their patients' families. The author would like to extend sincere appreciation to Charles Christopher Sheid, Coordinator of Marketing and Communications, Western Wyoming Community College, for his editorial expertise with this manuscript. Further, the author acknowledges, and expresses sincere appreciation to, the students, patients, families, physicians and others, who gave informed consent for the use of their statements and/ or images in this manuscript. Finally, the author thanks Patricia A. Kelly and the Kelly Family from Munster, Indiana, for their relentless involvement in this unique approach in medical education.

REFERENCES

AAMC. 1998. Learning objectives for medical student education—Guidelines for medical students. Medical School Objectives Project Report 1. Washington, DC.

Aziz MA, McKenzie JC, Wilson JS, Cowie RJ, Ayeni SA, Dunn BK. 2002. The human cadaver in the age of biomedical informatics. *Anat Rec* 269:20–32.

Boyle CJ. 2006. Fostering leadership and professionalism. *Am J Health Syst Pharm* 63:210, 212.

Bryan RE, Krych AJ, Carmichael SW, Viggiano TR, Pawlina W. 2005. Assessing professionalism in early medical education: Experience with peer evaluation and self-evaluation in the gross anatomy course. *Ann Acad Med Singapore* 34:486–491.

Devereux K, Foley S. 2014. A Parting Gift, Parts 1 and 2, Trinity University, vimeo.com/110253032https://vimeo.com/110253032> and vimeo.com/110557616<https://vimeo.com/110557616>, Dublin, Ireland.

Dyer GSM, Thorndike MEL. 2000. Quidne mortui vivos docent? The Evolving Purpose of Human Dissection in Medical Education. *Acad Med* 75:10:969–979.

Engel GL. 1977. The need for a new medical model: A challenge for biomedicine. *Science* 196:129–136.

Escobar-Poni B, Poni ES. 2006. The role of gross anatomy in promoting professionalism: A neglected opportunity! *Clin Anat* 19:461–467.

Flocke SA, Goodwin MA, Stange KC. 1998. The effect of a secondary patient on the family practice visit. *J Fam Pract* 46:429–434.

Forrest ME. 2008. On becoming a critically reflective practitioner. *Health Info Libr J* 25:229–232.

Giles B. 2015. "IUN Prosection Program Changes Course of Kouts Woman's Life," *Northwest Indiana Times*, 26 August.

Gunderman RB, Wilson PK. 2005. Viewpoint: exploring the human interior: The roles of cadaver dissection and radiologic imaging in teaching anatomy. *Acad Med* 80:745–749.

Hundert EM. 2001. A piece of my mind: A golden rule: Remember the gift. *JAMA* 286:648–650.

Jones TW, Lachman N, Pawlina W. 2014. Honoring our donors: A survey of memorial ceremonies in united states anatomy programs. *Anat Sci Educ* 7:219–233.

Kolb DA. 1984. *Experiential Learning.* Englewood Cliffs, NJ: Prentice Hall.

Lachman N, Pawlina W. 2006. Integrating professionalism in early medical education: The theory and application of reflective practice in the anatomy curriculum. *Clin Anat* 19:456–460.

Lewis-West B, Hiatt N. 2011. Keeping Humanity in Human Anatomy: Panel TV Show Debate and Discussion. Spirit & Place Festival: The Body Sound Medicine, WFYI TV and Radio, Indiana University School of Medicine.

Library of Medicine, National Museum of Health and Medicine. CD-ROM V2.0, September 2004.

Louw G, Eizenberg N, Carmichael SW. 2009. The place of anatomy in medical education: AMEE Guide no 41. *Med Teach* 31:373–386.

Pawlina W, Hromanik MJ, Milanese TR, Dierkhising R, Viggiano TR, Carmichael SW. 2006. Leadership and professionalism curriculum in the Gross Anatomy course. *Ann Acad Med Singapore* 35:609–614.

Pawlina W. 2006. Professionalism and anatomy: How do these two terms define our role? *Clin Anat* 19:391–392.

Reilly C. 2015. Gaining a body of knowledge. *The Medical Independent*, 2 April 2015.

Rizzolo LJ. 2002. Human dissection: An approach to interweaving the traditional and humanistic goals of medical education. *Anat Rec* 269:242–248.

Schon D. 1983. *The Reflective Practitioner, How Professionals Think in Actions.* New York: Basic Books.

Shapiro J, Talbot Y. 1991. Applying the concept of the reflective practitioner to understanding and teaching family medicine. *Fam Med* 23:450–456.

Štrkalj G. 2014. The emergence of humanistic anatomy. *Med Teach.* 36:912–913.

Sukol RB. 1995. Building on a tradition of ethical consideration of the dead. *Hum Pathol* 26:700–705.

Swartz WJ. 2006. Using gross anatomy to teach and assess professionalism in the first year of medical school. *Clin Anat* 19:437–441.

Talarico EF Jr, Buchler LT. 2011. The Case of the Dark Dot. MedEd PORTAL®, American Association of Medical Colleges.

Talarico EF Jr, Koveck SM. 2012. Mr. William's pneumonia. MedEd PORTAL®, American Association of Medical Colleges.

Talarico EF Jr, Prather AD, Hardt KD. 2008. A case of extensive hyperostosis frontalis interna in an 87-year-old female human cadaver. *Clin Anat* 21:259–268.

Talarico EF Jr, Prather AD. 2007. A piece of my mind. Connecting the dots to make a difference. *JAMA* 298:381–382.

Talarico EF Jr, Vlahu Ac. 2015. Characterization of kyphoscoliosis and associated giant hiatal hernia in a 97-year-old female cadaver. *Eur J Anat* 19(3):257–276.

Talarico EF Jr, Walker JJ. 2007. Introduction to the Cadaver Experience: Orientation to Human Gross Anatomy & Embryology. Washington, DC: Stand-alone, Permanent Exhibit, National Library of Medicine.

Talarico EF Jr. 2010. A human dissection training program at Indiana University School of Medicine-Northwest. *Anat Sci Educ* 3:77–82.

Talarico Jr EF, Frantz TL. 2013. Mrs. Wrubel's Wobbly Gait. MedED PORTAL®, American Association of Medical Colleges.

Talarico EF Jr, Hiemstra TM. 2013. A case of iniencephaly in a 36 week-old human female fetus. *Eur J Anat* 17(4):209–219.

Talarico EF Jr, Platt BL. 2014. My Beautiful Hair. MedED PORTAL®, American Association of Medical Colleges.

Talarico EF Jr. 2013. Change in paradigm: Giving back identity to donors in the anatomy laboratory. *Clin Anat* 26:2:161–172.

Terry M. 2014. Dear Joseph. *Pulse*, 21 February.

Trotman P, Nicholson H. 2009. Donated to Science. Dunedin, New Zealand: PRN Films, Port Chalmers.

Wagoner NE, Romero-O'Connell, JM. 2009. Privileged learning. *Anat Sci Ed* 2:47–48.

7

COMMEMORATION PRACTICES AT OTAGO: EXPERIENCES FROM A BICULTURAL SOCIETY

Natasha AMS Flack,* Kathryn McClea and Helen D. Nicholson

*Department of Anatomy, Otago School of Medical Sciences
University of Otago, Dunedin, New Zealand
* natasha.flack@anatomy.otago.ac.nz*

ABSTRACT

New Zealand is a bicultural country with the Treaty of Waitangi ensuring equal rights for Māori (the indigenous people) and *Pākehā* (non-Māori). The commemorations practised at the University of Otago reflect the cultural beliefs of Māori and the increasingly multicultural nature of New Zealand society.

In Māori culture, the dead body is recognized as *tapu* (sacred or restricted) and it is necessary for Māori students to take part in a ritual to allow them to enter and work in the dissecting room safely. The Department of Anatomy holds a clearing of the way ceremony (*whakawātea*) at the beginning of each academic year to meet this need. Initially, the *whakawātea* was held only for Māori students but it is now offered to all students studying Anatomy.

Interestingly, most students attend the ceremony, and this chapter explores the reasons why students choose to come to the *whakawātea*. Another commemoration is held towards the end of the academic year to remember the lives of those who have bequeathed their bodies. This student-led, thanksgiving service brings together the families of the donors and provides them with some closure. Families also have the opportunity to meet with students and understand the value of their family member's gift. Further commemoration processes occur following the completion of dissection.

INTRODUCTION

Commemorative events around the area of body donation are held at many medical schools and universities across the world (e.g., Xie, 2006; Lin *et al.*, 2009; Kooloos *et al.*, 2010; Sakai, 2008; Webb, 2010; Park *et al.*, 2011; Martyn *et al.*, 2013). They usually reflect the cultural norms or values of the society of the donor program, but they also share a similar central theme of respect and gratitude to the donors and their family and friends. They occur in various formats, at differing times during the dissection process, and where the student body has a strong religious affiliation the ceremonies often have a strong religious component. The services may involve just the students, or the donors' families as well as students. In some circumstances, the donors themselves are included prior to their death (Lin *et al.*, 2009).

In the case of services that only include the students, these are often held in the dissecting room environment and involve giving the name back to the anonymous donors as a way of remembering their personhood (Vora, 1998). In some universities, such as the Buddhist-based Tzu Chi College of Medicine in Taiwan, the families take part in a moving ceremony where they give, or hand over, their loved ones as "Great Teachers" to the students who will be learning from them. The students may also visit the donor beforehand as part of a dedication service (Lin *et al.*, 2009). Similarly, the body donors to Juntendo University in Japan assemble at the annual general meeting of the *Shiraume-Kai* (society of body donors to the Juntendo University) where they are formally thanked for their imminent gift. Second year medical students not only attend the meeting as part of their dissection curriculum, but they also are involved in the organization of the meeting and meet and greet with the donors (Sakai, 2008). In Korea, appreciation services are held at the beginning and end of dissection and in catholic-based medical schools, the students participate in mass before dissection, and then again afterwards with the donor' families (Park *et al.*, 2011). A permanent monument for donors' families, staff and student to visit, at any time to pay their respects, has been installed at the Radbound University Nijmengen Medical Centre in The Netherlands. An annual memorial service for families and students is also held in this location, with the monument playing the central focal point of the ceremony (Kooloos *et al.*, 2010).

In a number of Westernised universities, the content of such memorial services appears to be similar. The service is typically held at the end of the academic year and students and families of the donors are present, and may contribute to the event in a variety of ways (Xie, 2006; Webb, 2010). However, the format of a commemorative event may differ depending on not only the religious affiliations of the individual institutions, but also on whose needs they are addressing; the process of body donation affects many people including the donors' families and friends, students, administrative, technical, and academic staff.

The importance of the dissecting room in educating our health professional and science students is becoming more clear, not only in promoting learning of anatomy but also in developing team work, ethical and professional behaviors and helping students to cope with death (Lempp, 2005; Korf *et al.*, 2008; Bockers *et al.*, 2010). As a consequence, helping students to understand the gift that the donor has given and the role of the families in facilitating this gift is a key part of the learning.

In this chapter, we discuss the commemorative events that occur at Otago in a given academic year. Interestingly, the order of the events during the year also reflects their development, with the initial "clearing of the way" (*whakawātea*) ceremony being the first to be instigated. However, before discussing the ceremonies it is important to understand a little of the cultural background of New Zealand.

CULTURAL BACKGROUND OF NEW ZEALAND

New Zealand is a bicultural state, which recognizes the rights of the indigenous people, Māori, as well as the incoming immigrants (*Pākehā*). Māori travelled from Asia via Polynesia to *Aotearoa* (New Zealand) by boat and settled ~ 800 CE. Over the next 500 years, further canoes arrived adding to the establishing Māori society (Broughton, 1993). The society was, and still is, focused mainly on tribes (*iwi*), which can be traced back to the ancestors who came on the original seven canoes. Europeans, and other nations, came to New Zealand much later. Captain Cook "discovered" New Zealand in 1769 and increasing numbers of European migrants followed this. In 1840, the Treaty of Waitangi was signed between the Māori *iwi* and Great Britain. The Treaty enshrines the rights of the indigenous Māori and *Pākehā* and is based on reciprocal rights and responsibilities. Māori currently make up 15% of the population (Statistics New Zealand, 2013), although the numbers of Māori working as health professionals have historically been a much lower percentage. Facilitating the training of Māori is a key driver of the University of Otago and the New Zealand Government to address social, economic, and health inequities experienced by Māori. Equally important is ensuring that all health professionals are aware and respectful of *tikanga* (Māori protocol) and other cultures.

In particular, the cultural response to death and dying for Māori is special. Like many other cultural groups, the events surrounding times of serious illness, dying, death and grieving are among the most sacred and important in Māori life. They are steeped with *tapu* (sanctity) and *tikanga*. The rituals and customary practices are elaborate, the *reo* (language), *karakia* (incantations and prayers), *waiata* (chants and songs) and *pūrākau* (oral literature) are symbolic and poetic, and the responses and reactions by all are concerned, open and expressive (Ngata, 1987).

These cultural values/views are often present in our Māori students. Thus, in order to enter the dissecting room and subsequently undertake dissection in a safe way that is free of untoward fears, a ritual, *whakawātea*, occurs with support from our local Māori community at the beginning of each academic year.

WHAKAWĀTEA, "CLEARING OF THE WAY" CEREMONY

The *whakawātea* occurs in the Department of Anatomy at the University of Otago each year before the commencement of any teaching in the dissecting room. The first ceremony took place in 1989 at the request of several Māori students and after staff noticing that some Māori students were not attending the dissection classes. In this first year, only Māori students were invited to attend the ceremony. However, in the following year, Māori students were invited to bring a friend and 2 years later the ceremony was opened up to all students, regardless of their cultural background. Now, all University of Otago students studying anatomy are invited to attend the *whakawātea*. Attendance is a non-compulsory part of the orientation program for the medical and dental students. A single ceremony is dedicated to this specific group of individuals and turnout is very high with over 80% of the class attending this optional event. Other health professional and science students studying anatomy are invited to an additional, identical ceremony. Although there are fewer students per program that are present, the ceremony is still well attended and well received.

The ritual of the *whakawātea* recognizes and respects the complex nature of *te tapu o te tangata* (the sanctity of the person). A dead person (*tūpāpaku*) is considered *tapu* and people who are in contact with, or are in the presence of, a dead body also become *tapu*. The *whakawātea* ritual clears the space of any spiritual components that may impede learning so that students and staff can work in an unencumbered way.

At the time of the ceremony, students gather in a lecture theatre where representatives of the local *iwi*, Ngāi Tahu, the Office of Māori Development at the University, the Anatomy Department, the Associate Dean Māori, Māori teaching staff and the *kaumatua* (Māori elder) explain the background of the ceremony. It is firstly acknowledged that all cultures have their own perspectives on death and how they view the dead person. Then it is explained that the *whakawātea* provides an appropriate way for people, especially those with a Māori or Polynesian background, to enter the dissection room for the first time. It is not a religious service for the dead, but a ceremony for the living to allow them to study without fear.

The students are then led into the dissecting room by the *kaumatua* and Māori staff. The *kaumatua* begins by reciting *karakia* (incantations) as he walks through the lines of covered *tūpāpaku* (cadavers) with the students following him, and *waiata* (songs) are sung towards the end of the ceremony. At the end of the *whakawātea*, and after each session in the dissection room, students remove the *tapu* by cleansing with water;

washing their hands following health and safety procedures and Māori students may also sprinkle water over themselves to follow *tikanga*.

REASONS FOR ATTENDANCE AT THE *WHAKAWĀTEA*

The *whakawātea* is attended by a large number of students, who come along for a variety of reasons. The results of a survey of the students who attended, and their reasoning for why, have been recently published (Martyn *et al.*, 2013). The *whakawātea* provides a time for some students to reflect spiritually on the situation and 20% of our students in 2011 mentioned the purpose for their attendance was for spiritual reasons. Seeking permission from the donors to embark upon the dissection during the year seems to be another reason why students attend the *whakawātea* and perhaps reflects why the ceremony is well attended by our students. Furthermore, the ceremony allows all students to pay their respects and offer thanks to the donors before the process of dissection occurs. This is not always affiliated with religious belief but appears necessary for some students to go through before they are comfortable with the act of dissecting. The ceremony may also remind students that it was the donors' desire to be part of the bequest program and that they gave their permission to be there.

On a practical level, many students will attend in an attempt to prepare themselves for dissection, treating it as an informal orientation to the environment in which they will be working in, without the added pressure of having to perform or learn (Martyn *et al.*, 2013). From our data collected in 2011, it was found that most (92%) students found that the *whakawātea* helped them to feel more comfortable about entering the dissection room in the future. Similarly, some students will attend out of interest, with the intention of finding out what happens at the ceremony and the process around it (Martyn *et al.*, 2013).

THANKSGIVING SERVICE

While the *whakawātea* addresses the needs of the students, the thanksgiving service at Otago was developed initially to support the families of the donors. One of the main reasons for initiating a thanksgiving service was to help the donors' families through the grieving process. Through our conversations with local funeral directors we had become aware that some families were finding it particularly difficult to cope after their loved one came to the medical school. There were various reasons for this. A significant reason was that they did not have a normal funeral with the coffin present, and while it was possible for them to have a memorial service, many families declined this option and the lack of closure was causing difficulty in working through their grief.

Another concern raised was that they were not clear what happened to their loved ones after they came to the medical school and also the potential impact of the bequest on students. In response to this feedback, we held our first thanksgiving service in 2004 and have continued this annually since then. While the format of the service has developed over the years, the basic underlying principles have remained. The service is aimed at the donors' families and friends to allow them to remember the good times they had with their loved ones and to explain why the donations are so important to the training of health professional and science students. The service has a high degree of student participation and is inclusive and not aligned to any single religion.

The thanksgiving service is held each year, usually in September, which is towards the end of our academic year. While the University of Otago has its main campus in Dunedin, approximately 66% of its donors come from the Christchurch region. The University also has a campus in Christchurch where one-third of the medical students and other health professional students undertake their clinical studies. Because of this, the service is held in alternate years in Dunedin and Christchurch and the students in each locality contribute to the services in their respective areas. The service is held in the evening. In Dunedin, it usually takes place in the Town Hall complex, while in Christchurch it has been held at several different large public venues. Invitations are sent to the families whose loved ones have come to the medical school in the last two years. Families who have been bereaved in the three months prior to the date of service are invited to the service held in the following year. There is no limit on the number of the donor's family and friends who may attend and the families are encouraged to bring a photo of their loved one with them to the service. They are also invited to send in any memories of their loved ones that are read out by a student at the service. All students who take part in learning in the dissection room are invited to attend the service but there is no compulsion for them to attend. Students are also invited to take an active part in the service in a variety of roles e.g. ushering, reading, singing, and performing musical items. Prior to the thanksgiving service, students are asked to reflect on their time working with their cadaver and to put these reflections together in the form of a piece of writing, prose or poetry. Some of these are then read in the thanksgiving service.

On the evening of the service, families are greeted by student ushers who show them to seats and take the photo of their loved one to place it on a table at the front of the room. The service then proceeds according to the order of service (Fig. 1) and usually lasts about 90 minutes. The short address highlights the University's thanks for the bequests, and provides a brief outline of how the donations are used and the large number of students who benefit from these very special gifts. Towards the end of the service, families are invited to light a candle in remembrance of their loved ones. After the service, the families are encouraged to stay for refreshments, allowing the families to meet and talk to the students and staff. This is probably one of the most helpful times for the families. As the families leave they are given a rose in remembrance.

Following the death of their loved one, the Department of Anatomy also sends a condolence card and Reflections booklet to the family, to commemorate the act of the donation. The card acknowledges the sacrifice of the family in allowing their loved one's body to come to the medical school and the booklet contains words of thanks written by the students (Fig. 2).

We have not formally surveyed the families regarding their perceptions of the thanksgiving service, but the letters of appreciation that we receive, positive feedback from funeral directors, and the fact that so many families choose to attend suggest that the service meets a need. While the service is intended to be primarily for the families, our data suggest that the students that attend also find it a positive experience. In a recent survey of the students (of all ethnicities) who attended the service, 93% found it a helpful experience with 44% finding it very or extremely helpful. When asked why they attended the service, the most common reason given was "to show gratitude to the families" (69%).

University of Otago

The Department of Anatomy welcomes you to this

Service of Thanksgiving

Thursday 3 September, 2015, 7:00pm
The Glenroy Auditorium, Dunedin Centre

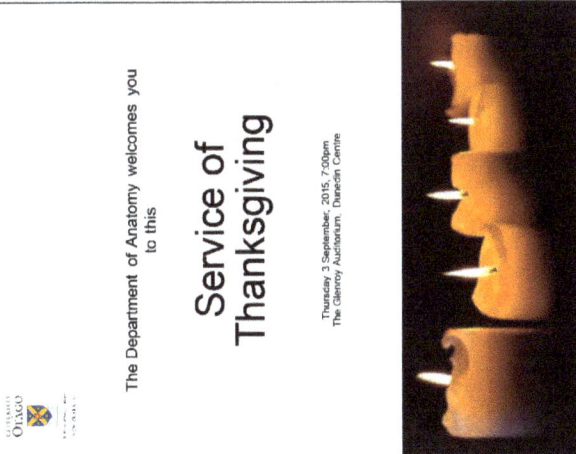

Bagpipes:	Medical Students
Waiata:	Te Oranga ki Ōtākou / Māori Medical Students Association - Otago
Mihi Whakatau:	Kaumatua
Gathering Prayer:	Chaplain, University of Otago
Family Messages:	Tribute messages from family members, read by students
Music:	Administrative Assistant, Department of Anatomy
Family Messages:	Tribute messages from family members, read by students
Music:	Health professional students
Family Messages:	Tribute messages from family members, read by students
Music:	Dental Students
Key Speaker:	Consultant Urologist, Dunedin Hospital
Candle Lighting:	Family and friends are invited to come forward and light a candle in memory of their loved one. Bequest Administrator and Professional Practice Fellow
	Music by 3rd year Medical Student
Student Messages:	Students' own words of thanks
Music:	Pacific Islands Health Professional Students' Association
Student Messages:	Students' own words of thanks
Song:	O GOD, OUR HELP IN AGES PAST Tune: St Anne O God, our help in ages past, Our hope for years to come, Our shelter from the stormy blast And our eternal home. The world in which we take our place Lies folded in your care; Love will remain and life persist And knowledge flourish here. Lord of the living and the dead, We honour those who gave To others of their very self To teach, inform and save. Their lives were fruitful of much good To their community, With gratitude we celebrate Their lasting legacy. We lift their names, O Lord, to you. Great guardian of each soul; True wisdom, everlasting love, In whom are all made whole. Verse 1 Isaac Watts (1719) Verse 2-5 Colin Gibson (2005)
Closing:	Chaplain, University of Otago

Please join us for light refreshments after the service, and accept a rose in memory of your loved one.
Don't forget to collect the photo of your loved one before you leave tonight.

Figure 1. Order of Service for the thanksgiving ceremony.

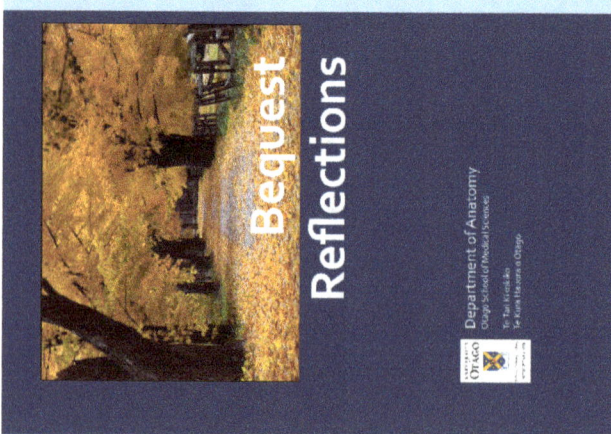

Figure 2. An example of a student's personal contribution provided within the Reflections booklet that each family receives at the thanks-giving service.

Table 1.   Thematic arrangement of some students' comments regarding the thanksgiving service.

Theme	Medical Students' Comments
Gratitude	It was weird seeing the photos of the cadavers, looking so alive. That was hard but nice too.
	Made me aware of how lucky we are and how each body has a really different family background... made me more determined to enjoy it
	Was a nice time to reflect on what an amazing thing it is for a person and their family for them to donate their body.
	It made me a bit sad but mostly very grateful that we were given this opportunity and could show our thanks by learning as much as we could.
More comfortable/ permission	It made me feel more comfortable about carrying out the dissections because I had talked to some families and heard their stories.
Professional development	Reconfirmed my belief that I am entering a profession which is respectful of others in both life and death.
Families	It was incredible to meet the families and talk to them. It was a huge gift from them as well, not just from the donor.
	So sad seeing families of the people we dissected, after viewing the bodies as a tool it was crazy to associate them with the people in the room. I hoped that they didn't see me as brutal after I had been chopping up their loved ones, but it was awesome to see how appreciative they were about it.
Personhood	It was quite emotional to hear the stories of the families, really brought it home that there is a lot more to the people on the table we work on

Other reasons given were "because it seemed the right thing to do" (21%) or "out of curiosity" (4%). Some students also added their own comments on what the service meant to them and some of these are shown in Table 1. The comments echo the themes of gratitude to the donor and helping to understand the impact that donating a body to science has on the donor's family, e.g. "I appreciated how difficult it was for the families of the donors."

Conversations with donor families about the desire that the donors had to give their body to the medical school allowed some students to feel like they were "given permission" to study the bodies, and inspired the students to study harder. This theme was also raised by students in the film "Donated to Science" (Trotman and Nicholson, 2009). Students also commented that the event re-emphasized that the cadaver had once been a person, not just a learning resource, e.g., "It was incredible to hear the different stories about the people who we were dissecting, it made them more real and the process of dissection more of a privilege."

AT THE END OF DISSECTION

It has been argued that students learn much more than anatomy from working with cadaveric material in the dissecting room. Professional development and, in particular, coming to terms with death are areas that the students face (Lempp, 2005; Korf *et al.*, 2008; Bockers *et al.*, 2010). At Otago, a module on death and dying runs alongside the students' time in the dissection room. During the making of the film "Donated to Science" (Trotman and Nicholson, 2009), we also became aware of the importance for some students of trying to put all of the body back together. All of the parts of one body have always been put into a single coffin for cremation, however, for some students putting the various body parts together to make a body that resembled the original cadaver was important so that they could say goodbye to their cadaver. Therefore, in the last session of their dissection in the third year, the medical students put the body back together the best that they can, including returning the internal organs. Then the body is covered up and the students take a moment of silence to reflect on their experiences with the cadaver.

Once the teaching is finalized for the year, the Māori *kaumatua* returns to formally farewell the donors' bodies in a final ceremony (*poroporoaki*). During this time, the donors are thanked for their gift and service, and farewelled to rest in peace. For those donors that have specified that the ashes be returned to their families they are given back. The ashes of the donors whose families have not requested their return are scattered at the local cemetery. The Department of Anatomy has a dedicated rose garden at the local cemetery commemorating the bequestees and their families who have generously donated their bodies for teaching and research.

CONCLUSIONS

The commemorative events at the University of Otago have developed substantially over the last three decades to become a unique and comprehensive collection of services. Not only are the donors and the families of the donors thanked for their gift to medical education and research, but also the students themselves are allowed the opportunity to reflect on the reality of the situation, show gratitude and gain permission from these unique individuals and give thanks to the donors for their invaluable gift. The initial service, *whakawātea*, allows the indigenous people of New Zealand to work in a culturally safe and unencumbered environment during their anatomy studies, while it gives others the chance to become accustomed to the dissection room environment. This "introduction" may alleviate some pre-dissection anxieties that often accompany the reality of being confronted with, and dissecting a dead body. The annual thanksgiving service brings together the families of the donors, with the students studying anatomy, and provides a chance for the two groups to appreciate the difficulties of body donation and dissection. The families are encouraged to celebrate the lives of their loved ones, and the event aims to provide some closure to those who are struggling with the sacrifices that are made when a donor gives their body to medical education and research. These practices are well adapted to the bicultural society of New Zealand and, as evidenced by feedback from donors' families and students, are well received and meet the needs of all of those involved.

ACKNOWLEDGEMENTS

The authors would like to thank Dr. Louise Par-Brownlee for reviewing the manuscript and her advice regarding Māori culture.

REFERENCES

Bockers A, Jerg-Bretzke L, Lamp C, Brinkmann A, Traue HC, Bockers TM. 2010. The gross anatomy course: An analysis of its importance. *Anat Sci Educ* 3:3–11.

Broughton J. 1993. Being Maori. *N Z Med J* 106:506–508.

Kooloos JG, Bolt S, van der Straaten J, Ruiter DJ. 2010. An altar in honor of the anatomical gift. *Anat Sci Educ* 3:323–325.

Korf H-W, Wicht H, Snipes RL, Timmermans J-P, Paulsen F, Rune G, Baumgart E. 2008. The dissection course — necessary and indispensible for teaching anatomy to medical students. *Ann Anat* 190:16–22.

Lempp HK. 2005. Perceptions of dissection by students in one medical school: Beyond learning about anatomy. A qualitative study. *Med Educ* 39:318–325.

Lin SC, Hsu J, Fan VY. 2009. "Silent virtuous teachers": Anatomical dissection in Taiwan. *BMJ* 339:b5001.

Martyn H, Barrett A, Broughton J, Trotman P, Nicholson HD. 2013. Exploring a medical rite of passage: A clearing of the way ceremony for the dissection room. *Focus Health Prof Educ* 15:43–53.

Ngata P. 1987. Death, dying and grief: A Maori perspective. In: *The Undiscovered Country: Customs of the Cultural and Ethnic Groups of New Zealand Concerning Death and Dying*. Wellington: New Zealand: Department of Health. p5.

Park JT, Jang Y, Park MS, Pae C, Park J, Hu KS, Park JS, Han SH, Koh KS, Kim HJ. 2011. The trend of body donation for education based on Korean social and religious culture. *Anat Sci Educ* 4:33–38.

Sakai T. 2008. Body donation: an act of love supporting anatomy education. *Japan Med Assoc J* 51:39–45.

Statistics New Zealand. 2013. 2013 Census — Major ethnic groups in New Zealand. URL: http://www.stats.govt.nz/Census/2013-census/profile-and-summary-reports/infographic-culture-identity.aspx [accessed 12 January, 2016]

Trotman P. (Producer), Nicholson, H. (Editor). 2009. *Donated to Science*. Dunedin, New Zeland: PRN Films.

Vora A. 1998. An anatomy memorial tribute: Fostering a humanistic practice of medicine. *J Palliat Med* 1:117–122.

Webb C. 2010. Students honor first patients through memorial service. The University of Texas Medical School at Houston, Houston, TX. URL: http://www.uth.tmc.edu/med/slidewhows/2007-cadaver/index.html [accessed 12 January, 2016]

Xie Y. 2006. Memorial service pays tribute to anatomy cadavers. USCF Today, The University of California, San Francisco, CA. URL: http://today.ucsf.edu/stories/memorial-service-pays-tribute-to-anatomy-cadavers/ [accessed 12 January, 2016]

8

COMMEMORATIONS AND MEMORIALS IN ANATOMY: TRIBUTE TO THE DONORS AND THE INDIGENT GIVERS

Beverley Kramer*,‡ and Graham J. Louw†

*School of Anatomical Sciences, Faculty of Health Sciences,
University of the Witwatersrand, Johannesburg, South Africa,
†Department of Human Biology, Faculty of Health Sciences,
University of Cape Town, Observatory, Cape Town, Western Cape, South Africa
‡Beverley.Kramer@wits.ac.za

ABSTRACT

The fundamental basis of teaching and learning human anatomy is the dissection of the human body. However, the first encounter with a cadaver will affect many Health Sciences students emotionally. The emotional progression of students as they face dissection is well documented. The question is how to deal with those feelings; how the approach to anatomy may help or hinder a student's ability to empathize with patients. The solution may be in a formal introduction to the cadaver, treating her/him as if she/he is the first patient, and further, by developing relationships with donors and the family of the donor.

In the 21st Century, it has become necessary to de-ritualize the dissection process in order to smooth the way for scholars to have access to positive knowledge. We discuss here ceremonies which have been designed to assist Health Sciences students in coping with the transition from life to death and in accepting the donor as a bountiful gift. In addition, we explore information which is provided to the students at their first encounter with the cadaver.

"And so I return to that directive of Human Anatomy, 'to cut.' Beyond wounding, perhaps deeper, to cut can be to create, as we do when we are children, cutting construction paper into shapes and gluing them together into less perfect, more meaningful patchwork wholes. We each carry with us images of the structures of the human body that we learned from 'Gertie.' Every future patient, every diagram in a book, will in some way always refer to her. She has become a mother, a progenitor, and a gift that extends ever outward."

Helena Winston, *A Medical Student-Cadaver Relationship*

INTRODUCTION

Dissection of the human body has been acknowledged for many decades as an important aspect in the training of health care professionals. Attitudes towards this process of teaching and learning have altered over time and this has led to Dyer and Thorndike (2000) proclaiming that "Changes in the culture of medicine have carried anatomy from a research science, to a training tool, to a hazing ritual and rite of passage, to a vehicle for ethical and moral education".

Vesalius, who published *De Humani Corporis Fabrica* in 1543, embodied the guiding intellectual values of the Renaissance Period (1400–1600), which was a time of increased interest in both science and humanism (Dyer and Thorndike, 2000). The human body became an object of study, useful in revealing interwoven truths about mankind, nature and the divine. Health Science institutions currently provide a paradoxical experience through dissection because students obtain knowledge about death in a manner that is quite unusual in the problem-based curricula which currently exist. This has not always been the case. Cadavers have tended to be regarded as objects, a resource which is used to study anatomy. The same duality existed during the Renaissance, in that illustrations from the time typically show a partially dissected corpse with the Professor delivering information while the students are crowding around, struggling to see. In the background, skeletons with scythes and appropriate motifs, always in Latin (Godeau, 2009), decorate the dissection hall. After the Renaissance Period, the anatomy lesson was gradually de-ritualized in order to provide access to positive knowledge. In modern times, death has been reintroduced as a symbolic dimension.

In the late 19th and early 20th Centuries, dissection was envisaged as a transformative event and students used black humor to defuse tensions surrounding this quasi-legal and ethically questionable activity (Warner and Edmonson, 2009). In the mid-20th century, the view changed to that of providing an opportunity for physicians-in-training to learn "detached concern" (Warner and Edmonson, 2009). In the 21st century, there has been a shift to a more humanistic approach. Students are introduced to the concepts of life and death which they are expected to deal with as professionals.

Almost every Health Sciences professional remembers their first encounter with a cadaver in the human anatomy dissection hall. While some studies have been carried out to understand the emotions experienced at the time of that exposure (Tschering *et al.*, 2000; Williams *et al.*, 2014), the occurrence for many is a life changing event, a rite of passage. From the available descriptions of how students are introduced to dissection, it appears that the first exposure may either occur in a step-wise fashion over a period of time (Tschernig *et al.*, 2000), while for others it is an abrupt initiation (see the University of the Witwatersrand experience below). Some students use mechanisms for coping by naming their cadaver (Williams *et al.*, 2014), while a growing trend amongst institutions is the use of ceremonies and memorials (Vora, 1998; Kao and Ha, 1999; Morris *et al.*, 2002; Elansary *et al.*, 2009; Pawlina *et al.*, 2011; Jones *et al.*, 2014) to assist students with what may be for most, a difficult experience. These ceremonies have been initiated to thank the donors. Students are encouraged to reflect on their experiences and in some cases, have also been encouraged to meet and correspond with the loved ones of the donors (O'Reilly, 2011). A fitting end to such a close relationship is a ceremony which has been introduced by some institutions (O'Reilly, 2011) at the end of the academic year where students are given the opportunity to show their deep gratitude for the bequest of the body for dissection purposes.

This discourse deals with the annual ceremonies developed over many years at the two oldest South African Institutions: The University of Cape Town (UCT) and the University of the Witwatersrand (Wits) in Johannesburg.

THE UCT EXPERIENCE

At this Institution, a student's first encounter with the concept of dying, death and dissection is in the second half of first year, which precedes the start of full body dissection. During this first session members of staff discuss the history of dissection and the role of dissection in modern curricula for students in the Health Sciences. The students also watch a video produced by the University of Otago, Dunedin, New Zealand, in which body donors are interviewed prior to their passing, and the students who dissect those very bodies are interviewed repeatedly over a year to track changes to their thoughts and feelings (Trotman and Nicholson, 2009). Family members of the donors are interviewed about their thoughts of the donation by their loved ones and are also invited to the ceremony at the end of the academic year. The advantage of seeing such a video for the students is the overall realization that they are part of a much larger community of students who are being exposed to dissection of the human body. As the majority of the class comprises school-leavers, who are about 19 years old, the timing of this session is important. The students need to be given the time to consult with their parents, elders in the village, religious leaders, former teachers and so forth, prior to returning to second year when full body dissection begins.

At the beginning of second year, a special introductory day is held on which the students are given in-depth information about the process and value of dissection. This session is immediately preceded by a discussion with a panel consisting of a diverse range of religious leaders who discuss cultural traditions and rituals around dying, death and the disposal of human remains. This panel discussion followed by a question and answer session for the students is arranged by staff working within the medical humanities. After this panel discussion, the students will learn more about the long history of dissection and the values that this incorporates.

Reflecting further on these values as time passes is an important learning activity for them all. During this introduction, the students are reminded that each cadaver lying on the dissection table was once alive, a sentient, loving human being. The changing demographic composition of the class of medical students has resulted in the institution's realization that this type of introduction is an essential feature of the learning process. For many students, the death of a loved one is followed by burial of an intact body, as soon as possible. This is the primary discourse of these students which is acquired in their home environment, during their schooling and is conveyed to them by religious leaders. When they are confronted by a box of human bones used to study osteology and then an embalmed body on a dissection table, this secondary discourse to which they are exposed in the dissection hall may clash with their primary discourse and may lead to an inability to move forward. As a department, we realized the importance of addressing this matter and thus of assisting our students to commence successfully with dissection without any hindrances in their paths. After the introductory talk, the students proceed to the dissection halls and "meet" their cadavers for the first time. In order to overcome their feelings about touching an embalmed body, the students complete a short surface anatomy practical session where they draw various landmarks on the skin and demonstrate the outlines of the organs.

At the end of the academic year, a special Dedication Ceremony is held which is a reflection on the journey undertaken during the year and a celebration of the act of giving. This is a non-religious event and is driven by the students, where they will sing, dance, read poetry, act out a cameo, play musical instruments, and use any other form of free expression in order to demonstrate their gratitude to the person who bequeathed their body to the department. Recently, we have begun inviting the family members of donors to this celebration. This has provided the students with an opportunity to learn the story behind the donor, and also to show family members how deeply respectful of the donor the students are. The students are able to express gratitude at benefitting from this wonderful gift. Family members express their relief at participating in the closure of the passing of their loved ones.

THE WITS EXPERIENCE

For decades, the honoring of cadavers through a "Dedication Ceremony" at the commencement of the year in the anatomy dissection halls, has been part of the anatomy course for all Health Sciences students entering into the anatomy department of the University of the Witwatersrand. The term "Dedication" was attached to this ceremony to indicate the dedication of the cadavers to the use of the students and the dedication of the students to learning from the cadavers. Over the years, this ceremony has come to mean much more. Professor Phillip V. Tobias, Head of the Department of Human Anatomy, initiated this ceremony in the early years of his tenure. The ceremony is held on the first day of the anatomy course in the second year of study. The reaction of the students to exposure to the cadavers is notably dissimilar in the different students entering the dissection hall. Some students appear to be calm, while others twitter and giggle in what is perhaps, their own way of dealing with fear. The majority of the students are silent in contrast to their usually talkative and expressive nature. Perhaps this is the moment of realization that the cadavers were living human beings up until very recently, that they were brothers/fathers/mothers/wives/partners, that someone still mourns their passing. Studies have in the recent past, addressed this emotional reaction of students and suggest that the topic of death must be dealt with and that students should be prepared in advance for the dissection experience (Tschernig *et al.*, 2000).

The dissection hall in which the ceremony is held is large and "sterile". It contains at least 30 stainless steel dissection tables with gleaming white plastic shrouds draped over the cadavers. The students (medical, dental, physiotherapy, occupational therapy, pharmacy, nursing, science and biomedical engineering) are provided with chairs between the cadavers and await the commencement of the Dedication Ceremony in silence. The senior academic staff of the Department and the Faculty (the Dean, Heads of other Departments, the Inspector of Anatomy from the Provincial Government) enter the hall in full academic regalia. As in the Cape Town experience, the history of dissection is delivered, informing students of the early years of anatomy and the difficulties in obtaining human cadavers for dissection. The National Health Act number 61 of 2003 of South Africa is referred to at this ceremony. It is made evident to students that in the recent past, legal mechanisms for obtaining human bodies have been put in place in countries around the world. This insightful talk was for many years followed by a brief non-denominational service by a religious leader. Finally, the Dean welcomes the students and impresses on them the privilege of dissecting human cadavers and having sufficient cadavers to allow all students to dissect. At Wits, the majority of cadavers since the inception of the anatomy program in 1921, have come from unclaimed cadavers, but in recent years, this has changed to an increasing number of bequeathed cadavers (Kramer and Hutchinson, 2015).

Over the years and with changes in the demography of the students, the use of a clergyman from different religions was introduced. As the classes of students were starting to change, to become far more representative of the different cultures in our society, we utilized a Christian clergyman of different denominations (e.g., Methodist, Roman Catholic, Anglican) one year, an Islamic moola the next, a Jewish rabbi the following year, and so on, in order to show respect for both the students and the cadavers who came from a variety of different groups.

A closing or "burial" ceremony was introduced in more recent years following completion of the dissection course. Staff and students once again gather in the main dissection hall. The nature of the closing ceremony is one of thanks to the cadavers for their contribution to the teaching of the students. The students are thus able to say an informal goodbye to their cadaver before burial, although unfortunately this has not yet been in the form of poems, essays or tributes as presented by students at Yale (Morris *et al.*, 2002; Elansary *et al.*, 2009) and in Cape Town (see the UCT experience).

In more recent years there have been changes to the ceremony in our Department. A religious leader is no longer included in the ceremony, as it was felt by some staff that students subscribe to a variety of different religions. Thus, having one clergyman, even if they are asked to provide a non-denominational service, is now perceived as not acceptable. However, as some students may wish to speak about death and dying, it may be valuable to have clergymen available (Tschering *et al.*, 2000).

My (BK) personal experience with my first cadaver as a student following our Wits dedication ceremony was one of huge gratitude, astonishment and awe. I was fascinated by the intricacies of the human body from the moment I started dissecting and that fascination has never left me. My first cadaver was my greatest teacher and I will always remember her gift with fondness and pride.

EMOTIONAL EXPERIENCE

While the exposure to a human cadaver for the first time may be a disconcerting experience in itself, the act of disarticulating, dismantling and depersonalizing the individual may also have an enormous emotional impact (Montross, 2008). Vora (1998) maintains that the memorial service is a first step in the emotional grounding required by students as they prepare for empathetic patient care. It is proposed that if one is well-prepared emotionally for the initial encounter with the cadaver then the students will also be more prepared for the variety of human emotions they will face throughout their careers (Vora, 1998). Thus the emotions of the students should be carefully monitored and discussed by the custodians of the cadavers, the anatomy lecturers.

In general, medical students make use of a variety of psychological coping mechanisms throughout the process of dissection (Williams *et al.*, 2014). Examples of these mechanisms are well-known such as joking, providing distasteful nicknames (see below), and even removing body parts from the laboratory (Williams *et al.*, 2014). Some students pull pranks to scare their classmates (Hafferty, 1988, 1991). This is said to represent a gallows humor (Freud, 1960; Bennett, 2003). It does, however, for some, infuse a measure of wit and comedy into the emotionally difficult task of dissection (Williams *et al.*, 2014).

While some departments do reveal the birth name of the cadaver, students will rarely use these names and instead, they themselves provide a name for the cadaver (Williams *et al.*, 2014). It is important to recognize that the practice of naming cadavers still persists amongst our students. Some staff and students consider this to be disrespectful, but students do remember these names for the rest of their professional lives (Williams *et al.*, 2014). The choice of such names may relate to a variety of personal features such as references to age, physical characteristics, a nickname, showing respect by using a formal address, reference to popular culture, naming cadavers with tattoos, by guessing their occupation, for philosophical reasons, and often using the cause of death

(Williams *et al.*, 2014). There is a segment of the student body who give reasons for not naming the cadaver, i.e. that they were not inspired to provide a name (45% of this group of students who were surveyed), the name was viewed as strongly disrespectful (20%) or mildly disrespectful (15%), or else it was attempted but the name was not durable (20%) (Williams *et al.*, 2014).

DISCUSSION

The encounter with a cadaver will affect many medical students emotionally (O'Reilly, 2011). This emotional progression of the students over time is well documented (Lella and Pawluch, 1988; Francis and Lewis, 2001; Swartz, 2006; Arráez-Aybar *et al.*, 2008; Redwood and Townsend, 2011). Most students approach dissection with feelings of shock and discomfort but towards the end of the course feel far more comfortable with the experience (Williams *et al.*, 2014). One can therefore argue that dissection is one of the ways to equip medical students with the emotional skills they will need when working with patients (Charlton *et al.*, 1994; Williams *et al.*, 2014).

The important question therefore, is how students may learn to deal with their emotions on working with their first cadaver and how this can help or hinder the student's ability to empathize with patients (O'Reilly, 2011). The solution may be in getting to know the cadaver, treating her or him as if they are the first patient, and by developing relationships with donors and donor's families (O'Reilly, 2011). This experience may well teach the students qualities that are essential for an holistic approach to patients, such as professionalism, respect and empathy — important qualities that are difficult to teach elsewhere in the medical curriculum (O'Reilly, 2011). When addressing professionalism, one can argue that respect for cadavers and the large amount of time spent working on dissection with tutors and peers may suggest that the anatomy laboratory is the first site for evaluating the professionalism of the student (Williams *et al.*, 2014). Thus, in order to support our students in their exposure to death, it is important for educators to provide the scaffolding for a carefully constructed introduction to the history of dissection and careful management of the progression of the student's personal development over the time spent in the dissection hall. In addition, offering the opportunity of a respectful closure to this phase of the process in the student's education by facilitating ceremonies to celebrate the selfless act of donation may assist with the emotional difficulties that students face during this period of their training.

RECOMMENDATIONS FOR THE FUTURE

There has been reflection on the value of anonymity of the cadaver compared to the provision of certain facts about the identity of the donor (Vora, 1998). Students are said to experience an improved academic performance when they are able to alter their relationship with a cadaver by obtaining additional relevant personal information (Talarico, 2013).

In order to ensure anonymity of a donor, a department may decide to withhold the birth name, age and cause of death (Williams *et al.*, 2014). However, some departments do reveal information about the cadaver, such as the age and cause of death, sometimes the occupation, but no birth name. This is said to result in a closer relationship with the body donor (Williams *et al.*, 2014). Acknowledgement of the personhood of the cadaver is said to be important (Bohl *et al.*, 2011; Crow *et al.*, 2012; Maron, 2012). This may be achieved by revealing the birth name of the donor, providing students with an opportunity to meet with donor's family (Penney, 1985; Winklemann and Güldner, 2004), and even offering students an opportunity to view a recorded interview with the donor before death (Bohl *et al.*, 2011; Crow *et al.*, 2012; Maron, 2012).

Other approaches have also been explored, such as using the term "donor", which has a more positive connotation than "corpse" or "cadaver" (Weeks *et al.*, 1995). In Thailand, cadavers are regarded as "ajarn yai", great teachers having the highest status in society (Penney, 1985; Winklemann and Güldner, 2004).

To respect corpses is to admit their humanity. This can be done through providing the student with a series of facts that aim to give the cadaver a social identity, namely their age, their medical history, the cause of death, and even their life story (Williams *et al.*, 2014). What has become clear is that providing personal information about a cadaver assists students in remembering that these people have lived (Godeau,

2009). To emphasize this, students could be asked to give their cadaver a name, mark them with their initials (Segal, 1988), or even write a biographic piece about the cadaver which they are dissecting (Godeau, 2009).

While there may still not be consensus in the literature about the mechanism of introducing students to the cadaver or the emotional impact of this on young professionals, the resounding sentiments which are disclosed by many health sciences professionals who have experienced human dissection is the privilege and the great gift which they have been given (personal communication, Kramer and Louw). Many refer to the deepest respect which they have for the donor (Kao and Ha, 1999).

REFERENCES

Arráez-Aybar LA, Castaño G, Casado-Morales MI. 2008. Dissection as a modulator of emotional attitudes and reactions of future health professionals. *Med Educ* 42:563–71.

Bennett HJ. 2003. Humor in medicine. *South Med* J 96:1257–1261.

Bohl M, Bosch P, Hildebrandt S. 2011. Medical students' perceptions of the body donor as a "first patient" or "teacher": A pilot study. *Anat Sci Educ* 4:208–213.

Charlton R, Dovey SM, Jones DG, Blunt A. 1994. Effects of cadaver dissection on the attitudes of medical students. *Med Educ* 28, 290–295.

Crow SM, O'Donoghue D, Jerry B, Vannatta JB, Britta M. Thompson BM. 2012. Meeting the family: Promoting humanism in gross anatomy. *Teach Learn Med* 24:49–54.

Dyer GS, Thorndike ME. 2000. *Quidne mortui vivos docent?* The evolving purpose of human dissection in medical education. *Acad Med* 75:969–979.

Elansary M, Goldberg B, Qian T, Rizzolo, LJ. 2009. The 2008 Anatomy ceremony: Essays. *Yale J Biol Med* 82:37–40.

Francis NR, Lewis W. 2001. What price dissection? Dissection literally dissected. *Med Humanit* 27:2–9.

Freud S. 1960. *Jokes and Their Relationship to the Unconscious*. New York: WW Norton.

Godeau E. 2009. Dissecting cadavers: learning anatomy or a rite of passage? Hektoen International. http://www.hektoeninternational.org [accessed 15 May 2015].

Hafferty FW. 1988. Cadaver stories and the emotional socialization of medical students. *J Health Hum Behav* 29:344–356.

Hafferty FW. 1991. *Into the Valley: Death and the Socilaization of Medical Students*. New Haven and London: Yale University Press.

Jones TW, Lachman N, Pawlina W. 2014. Honouring the donors: a survey of memorial ceremonies in United States anatomy programs. *Anat Sci Educ* 7:219–223.

Kao T, Ha H. 1999. Anatomy cadaver ceremonies in Taiwan. *Zhonghua Yi Shi Za Zhi* 29:175–177.

Kramer B, Hutchinson E. 2015. Transformation of a cadaver population: Analysis of a South African cadaver program, 1921–2013. *Anat Sci Educ* 8:445–451.

Lella JW, Pawluch D. 1988. Medical students and the cadaver in social and cultural context. In: Lock DG (ed.), *Biomedicine Examined*. Dordrecht: Kluwer Academic Publishers.

Maron DF. 2012. Why some medical students are learning their cadavers' names. www.healthland.time.com [accessed 15 May 2015].

Montross C. 2008. *Body of Work: Meditations on Mortality from the Human Anatomy Lab*. New York: Penguin Books.

Morris K, Turell MB, Ahmed S, Ghazi A, Vora S, Lane M, Entigar LD 2002. The 2003 anatomy ceremony: a service of gratitude. *Yale J Biol Med* 75:323–329.

O'Reilly KB. 2011. Humanizing anatomy: a medical student's first patient. www.amednews.com [accessed 5 April 2015].

Pawlina W, Hammer RR, Strauss JD, Heath SG, Zhao KD, Sahota S, Regnier TD, Freshwater DR, Feeley MA. 2011. The hand that gives the rose. *Mayo Clin Proc* 86:139–144.

Penney JC. 1985. Reactions of medical students to dissection. *J Med Educ* 60:58–60.

Redwood CJ, Townsend GC. 2011. The dead center of the dental curriculum: changing attitudes of dental students during dissection. *J Dent Educ* 75:1333–44.

Segal DA. 1988. A patient so dead: American students and their cadavers. *Anthropol Quart* 61:17–25.

Swartz WJ. 2006. Using gross anatomy to teach and assess professionalism in the first year of medical school. *Clin Anat* 19:437–441.

Talarico EF Jr. 2013. A change in paradigm: giving back identity to donors in the anatomy laboratory. *Clin Anat* 26:161–172.

Trotman P. (Producer), Nicholson, H. (Editor). 2009. *Donated to Science*. Dunedin, New Zeland: PRN Films.

Tschering T, Schlaud M, Pabst R. 2000. Emotional reactions of medical students to dissecting human bodies: A conceptual approach and its evaluation. *Anat Rec* 261:11–13.

Vora, A. 1998. An anatomy memorial tribute: Fostering a humanistic practice of medicine. *J Palliat Med* 1:117–22.

Warner JH, Edmonson JM. 2009. *Dissection — Photographs of a Rite of Passage in American Medicine* 1880–1930. New York: Blast Books.

Weeks SE, Harris EE, Kinzey WG. 1995. Human gross anatomy: A crucial time to encourage respect and compassion in students. *Clin Anat* 8:69–79.

Williams AD, Greenwald EE, Soricelli RL, DePace DM. 2014. Medical Students reactions to anatomic dissection and the phenomenon of cadaver naming. *Anat Sci Educ* 7:169–180.

Winkelmann A, Güldner FH. 2004. Cadavers as teachers: the dissecting room experience in Thailand. *BMJ* 329:1455–1457.

Winston H. 2012. A medical student-cadaver relationship. *Virtual mentor* 14(5): 419–421.

9

COMMEMORATIONS AND MEMORIALS IN CHINESE BODY DONATION PROGRAMS

Luqing Zhang and Jiong Ding*

Department of Human Anatomy, Nanjing Medical University, Nanjing, China
*dingjiong@njmu.edu.cn

ABSTRACT

Over the last 20 years, body donation programs have been initiated in a few large cities in China. Most of these while successful, have also met with difficulties. A major issue is that the number of donated bodies is far below what is needed for medical education in China. The reasons for the low number of bequests is multifactorial and due in part to the influence of traditional spiritual beliefs, poor legislation, low social recognition, poor collaboration between donor organisations, and the lack of effective processes for donation. Overcoming these obstacles will require that all sectors of the Chinese society cooperate. Various ways to commemorate cadaver donors have emerged and evolved in the regions where donation is successful. These are now being promoted gradually throughout the country. Donors are commemorated via body donation monuments, memorial parks, public memorial activities, online memorial platforms and memorial activities included as part of delivering anatomy teaching. These commemorations and memorials will actively help change traditional Chinese ideas, improve social recognition, forge a powerful bond of trust between the anatomists and medical students with the general community. It is envisaged that in the long term these practices will result in a greater awareness of humanistic side of medicine in students, reduce the tensions between doctors and patients and promote body donation programs in China.

INTRODUCTION

While cadaveric donor programs have been established in many parts of China over the last 20 years, they have been challenged by small donor numbers that are insufficient to meet the needs of medical education (Zhang *et al.*, 2008). The main obstacles to increasing the number of donors are cultural norms, inadequate laws and poor social recognition (Zhang *et al.*, 2008; Chen *et al.*, 2014). Commemorating donors is a feature of regions with successful donor programs, and adopting these practices throughout the country is an important part of improving donor numbers. These commemorations play a critical role in changing the traditional ideologies, improving the social acceptance, building strong links between anatomists and medical students with the general public, enhancing the humanistic attributes of medical students, and promoting the establishment of successful donor programs. The commemorations include the creation of the body donation monument (forest), public memorial activities, online memorial platforms and commemorative activities in anatomy teaching.

THE CURRENT SITUATION OF BODY DONATION IN CHINA

Cadavers are indispensable to the teaching of anatomy, however many Chinese medical schools lack sufficient donor numbers and, as reported in the general media, are at risk of closing their programs. To compensate, many anatomy educators have resorted to using PowerPoint slides with images, pictures, anatomy models, virtual anatomy software and bottled dissected specimens (Liu *et al.*, 2011). This inevitably diminishes the quality of teaching and learning in these medicine programs, and has impacted students clinical skills and readiness for practice.

Anatomy has a long history in China. The practice of study by cadaveric dissection can be traced to as early as 500 BCE, and is recorded in the first classic Chinese medical book, *Neijing*; *the Internal Canon of Medicine* (Bertschinger, 2008). Following this, there are records of dissection during the Wang Mang period of the Han Dynasty by the imperial physician (Myers *et al.*, 1954), and a few hundred years ago in the Song Dynasty, by the acupuncturist Wang Weiyi and the forensic medical examiner, Song Ci who is also regarded as the Chinese Father of Forensic Science (Goldschmidt, 2009). During this time, China was regarded as a leader in the anatomical sciences, but with the subsequent development of traditional Chinese medicine the anatomical sciences was gradually ignored. Under the influence of Confucian philosophy, the focus shifted away from the sciences to loyalty and filial piety, and established a medical ethic associated with benevolence, and prohibited human dissection. The dissection of the human body came to be regarded as treason and heresy, and was regulated by the feudal governments for many generations.

Western medicine was introduced into China only in the last hundred years. There however, has been a lack of cadavers for teaching at medical colleges since the early years of the Republic of China. A recent survey demonstrated a shortage of cadaveric specimens in more than 630 medical colleges and universities impacting over 1.76 million students. Currently, approximately 50,000 potential donors register annually in donor programs resulting in about 3,000 bodies donated per year. This number is far below what is needed for medical education. While some medical schools are able to provide one cadaver for every 6–10 students, more commonly 15–20 or more students share a cadaver. This shortage of cadavers is seriously impacting medical training particularly in clinical and surgical skills, and clinical diagnosis and treatment in China.

FACTORS AFFECTING BODY DONATION IN CHINA

There are many factors that affect the attitude of body donation in China including a poor understanding of the body donor programs and its social significance, inherent traditional norms, religious beliefs and customs, social values and stigma, pressure from families, regulation and management of donor programs including the process for donation, ethical and moral practices, personal and psychological factors, etc.

China has 56 ethnic groups, each having its own religion and associated ways to treat the human body and to commemorate the dead. The Han nationality, which accounts for 90.56% of China's population, is especially influenced by the traditional culture mainstream Confucian thought. This advocates "living is happy and death is sad", and promotes "real-life enjoyment" with great optimism for living. This is seen in Confucian philosophy that urges people to "cultivate moral character", "regulate family", "develop the country" and "harmonize the world". At the same time, it considers death sad and sets up hierarchies for burial and sacrifice to deliver a deep grief for death. Its viewpoint that "filial piety is to recognize that body, hair and skin are inherited from the parents, thus should not be damaged" makes body dissection and donation a prohibited area (Legge, 2004). In this framework of thinking, autopsy has always been regarded as treason and heresy. "When alive, the body should be intact and when dead, the body should be elaborately buried". This has been a deeply rooted concept in most Chinese people. People lack a balanced view on life and death, fear death, refuse to talk about death, and as consequence cannot make arrangements for death while still alive. Such ideas are widespread even in the circle of the well-educated young people. As a result, the elderly may have the willingness to donate their bodeis, but their children and relatives, as executors of their estate, may consider the donation "unfilial and outrageous behavior", and refuse to donate the body.

CHINESE COMMEMORATIONS FOR BODY DONATION

To address the dilemma this presents to medical education and to solve the problem of cadaver shortage, body donor programs have been established in the last 20 years. So far, the program is under way in more than fifty cities in China and is promising to provide a cornerstone for the future development of the medical sciences.

In this century, various types of memorials for donors have been established in China, including the donation monument (forest), public memorial activities, online memorial platforms, and pre-teaching commemorative activities.

Donation Monument

Since 2001, the Red Cross Society has set up donation monuments and memorial parks with engraved donor names in Nanjing, Shanghai, Beijing, Tianjin, and Shandong, Shanxi, Guangdong, Jiangxi and other provinces (city). The aim of initiative is to provide donor families with mourning places commemorating the benevolence of the donors to future generations, and to educate the society as a whole into changing its views towards donation. In September 2001, the first memorial monument, the Nanjing Voluntary Body Donor Memorial Monument, was built in Red View Park in the Qixia Mountain of Nanjing. On the monument is engraved "Birth and death are natural to human; dawn and the sunset glow both illuminate the world". In March 2002, a donor monument was built in the Fushou Garden of Shanghai by the Red Cross bearing the engraving "Life is valuable because it is short and one-time. The people listed here have overcome the limit of their life span and become eternal because they have sublimated the value of their life". Subsequently, further monuments were erected and include: the Body Voluntary Donation Monument in Evergreen Park in Beijing (April 2004), the Donation Monument Garden in Tianjin (May 2004), "Light of life" in Shimen Peak Cemetery in Hongshan District of Wuhan (March 2005), monuments in Wuhan and Anhui province (March 2005), Zhengguowanan Cemetery in Zengcheng Guangdong Province (May 2008), the first donation monument and theme square built for Inner Mongolia's body donors in Hohhot (August 2008), and in the Shandong province in Fushou garden in Changqing District of Jinan (December 2008). This trend has continued and in the last 5 years many more cities across the country have built memorial facilities with public memorial services organized by the Red Cross commemorating past donors with participation of enlisted donors and their family, staff responsible for donor programs and representatives from medical schools.

Donor Memorial Hall

In order to commemorate the donor's benevolence, promote body donation and provide memorial facilities for donor relatives, the Red Cross Society of various regions of China has established memorial halls for the donors. These include the first Red Cross Donor Memorial Hall built in the Fushou garden of Shanghai, September 2003 and the Wuhan Donor Memorial Hall officially completed in March 2007. These memorial halls commemorate the donor's life with displays of photos, paintings and photographs alongside texts providing the context of the donor's experiences, expertise and outstanding achievements.

At the same time, exhibition pavilions were set up in some medical schools for ethics education. A sound awareness of medical ethics significantly contributes to a high ethical standard in medical personnel.

In 2002, Tianjin Medical University set up "The Room of the Meaning of Life", that in 2009 was approved to be a patriotism education base by the Tianjin Municipal Government. The room was relocated and expanded in 2014 and used in medical ethics teaching. Students were expected to visit the room and talk to the family members of the donors to improve their patient-doctor skills. In this way, the medical ethics teaching extended beyond the classroom and enhanced students' awareness of the humanities.

In 2004, Nanjing Medical University established a medical ethics education center, which was relocated and expanded in 2014. The center showcases the eighteen-year (1996–2014) course of body donation in the city with "benevolence", "honesty", "education" and "aspiration" as the main lines of the message representing the great love and principles of the donors. At the center, professors and students together show their respect and gratitude for the dedication of the donors. This helps to improve the humanitarian and professional quality of medical students and plays an effective role in education at medical colleges and universities. The education museum (center) is not only the main window for foreign exchange and a medical humanity education base, but also a compulsory part of the human anatomy classes. It contributes to the students understanding of the spirit of selfless dedication of the donors and cultivates love for life while acquiring medical knowledge. Since the opening, the center has been visited by more than 30,000 students, parents of students, "donor friends" and experts from dozens of medical colleges and universities, and is highly appreciated. "Health care needs the ethics most, the ethics is most concentrated in medical education", said Chen Ying, the former Head of the Chinese Ethics Society, when expressing affirmation and praise to the center.

In August 2011, the Health Science College of Beijing University built a special wall — a memorial wall for body donors, where rows of crystal boxes were displayed with the donors' photos, names and dates of birth and death. For the donors, the wall is a memorial; for the teachers and students who benefit, the wall shows their gratitude.

In October 2014, a Donation Memorial Hall was officially opened to the public at Dalian Medical University, and became the base of humanistic education for its students. This is the first memorial hall for body donors in the Liaoning Province and includes a solemn Pay Room with the photos of the 206 donors surrounded by chrysanthemums. The hall commemorates the donors returning their most valuable bodies to society as their last contribution to humanity through the medical sciences. They are honored as the "silent teacher" by the medical students.

Body Donation Memorial Website

To permanently preserve the voice and image of the donors who have donated their bodies and to allow the relatives and friends to pay their tributes with no restriction of time and space, the largest memorial website for body donors (http://hongshizi.netor.com) was set up by the Red Cross Society of Shanghai in 2006. Since then, memorial websites were built for Anhui, Nanjing, Chengdu, Jinan, Qingdao and other cities by the Red Cross. A virtual space is created for each realized donor with her/his personal pictures, biography and album, and with interactive areas to allow the family members or friends to pay their respects with flowers, incense, songs and so on.

In 2008, the donation memorial website for Houde Park (http://hdy.njmu.edu.cn) went live at Nanjing Medical University (Zhang *et al.*, 2014). It was the first public welfare website in the Chinese medical colleges to systemically commemorate and promote voluntary body donation. The website includes student activities, Nanjing donor friend activities, news center, an explanation of the donation process and regulations, and an online memorial. The "online memorial" is a virtual memorial hall, where friends and family members of the donors can log into the corresponding memorial hall with a username and a password. After logging in, various memorial activities can be selected to commemorate the donor, such as giving flowers and songs. It also allows the donor's lifetime data such as their voices, images and memorial anthologies, to be uploaded. Thus far, more than 400 memorial halls have been created in the website, which have become very important platforms for the donor families to commemorate their loved ones.

The memorial sites also provide convenient platform for future donations. For medical education, they help address the problem of cadaver shortages and promotes the development of medical education. For medical students, the websites are a platform for moral and medical ethics education. For the family members and friends of the donors, they provide alternative and everlasting ways to commemorate loved ones. Memorial websites will have a profound and long-lasting impact on teaching, social, cultural and other aspects.

Gratitude Education in Anatomy Teaching

In most Chinese medical schools, anatomy teaching is carried out together with medical humanities education to let the anatomy teachers and students recognize the importance of body donation and to be respectful of the cadaver in their study of it. Teachers and students in the medical school are conscious to express their respect for the body donors in a variety of ways. The gratitude education is performed in various ways such as the cultivation of medical humanistic spirit and professionalism, standardized use of the body in anatomy class, thanksgiving activities for the donors, visit to the memorial exhibition related to body donation, and silent pay tribute to the body in anatomy class.

What the donors donate are not just their bodies, but more importantly, their great spirit of humanity, love and dedication. It takes great courage and noble love to make this selfless dedication for medical education and research. At Nanjing Medical University, body donation is used as a focus to deliver humanistic and quality education for students reflecting the atmosphere of the campus culture. Gratitude to donors is highly valued. At the University, the first systematic anatomy class is on gratitude education; the first local anatomy practical class is a silent tribute to the body. The gratitude education is incorporated into the whole process of anatomy teaching. Every effort is made to make every medical student understand that the relationship between the donor and student is not simply the relationship between anatomical specimens and users, but the relationship between contributors and beneficiaries, essentially the relationship between people. Anatomy is completed with the donor and learner as a whole and as individuals. It is important to have a deep understanding of the nature of the relationship between the donor and the medical students, and to recognize that the body donation is a noble act, so that students will deeply respect the donors and their life and learn human care to the entire body donation process and to the anatomy activities.

At the same time, the activities related to body donation are merged into the campus culture. The freshmen are organized to visit the thanksgiving exhibition for donors; student commentators are trained to introduce body donation and the moving deeds of the donors. Ideas of respecting life, optimism, dedication to the society are encouraged. The student representatives are encouraged to participate in the farewell to experience the noble charity of the donors, thus providing direct humanistic education for students.

Students are organized to participate in various social practices by the Communist Youth League, the student offices, schools' Red Cross, colleges and Department of Human Anatomy. Student volunteers visit the communities, donor families, donor volunteer memorial monuments to promote body donation, angular membrane and bone marrow donation, and to respect, serve and commemorate donors.

CONCLUSION

Over the past 20 years, China has begun to advance and promote body donation, and the relevant agencies continue to explore, develop and improve ways to commemorate the donors through an established multidimensional approach using both real and virtual worlds to commemorate the donors. They have paved the way for the public to break the shackles of traditional ideas and promoted body donation as a noble act. This has great and far reaching impact on changing traditional ideas, improving the social recognition, building strong trust among the anatomists and medical students with the general public. It will also help to enhance a humanistic quality of medical students and alleviate the tensions between doctors and patients in China.

REFERENCES

Bertschinger R. 2008. Huangdi Neijing. In: Selin H (Editor) *Encyclopaedia of the History of Science, Technology, and Medicine in Non-Western Cultures.* Berlin, Heidelberg, New York: Springer, pp. 1080–1081.

Chen D, Li J, Huang J, Pan A. 2014. Problems and countermeasure analysis of the body donation work in our country basing on comparison to that in foreign countries. *Chin J Clin Anat* 32:626–629.

Goldschmidt A. 2009. *The Evolution of Chinese Medicine: Song Dynasty, 960–1200.* London, New York: Routledge.

Legge J. *The Hsiao King or Classic of Filial Piety.* Whitefish: Kessinger Publishing, LLC. 2004.

Liu J, Zhong G, Zhao Z, Wang X, Wang J. 2011. Discussion and practice of the unpaid corpses donation to medical colleges. *Chin Med Ethics* 24:470–471.

Myers RH, Bray F, Tsuen-Hsuin T. 1954. *Science and Civilisation in China.* Cambridge: Cambridge University Press.

Zhang A, Cui Y, Wu W. 2009. Analysis on causes of cadaver donation obstacles and countermeasures. Chin Med Ethics 2:101–102, 147.

Zhang L, Wang Y, Xiao M, Han Q, Ding J. 2008. An ethical solution to the challenges in teaching anatomy with dissection in the Chinese culture. *Anat Sci Educ* 1:56–59.

Zhang L, Xiao M, Gu M, Zhang Y, Jin J, Ding J. 2014. An overview of the roles and responsibilities of Chinese medical colleges in body donation programs. *Anat Sci Educ* 7:312–320.

10

THE CEREMONY TO HONOR THE BODY DONOR AS PART OF AN ANATOMY OUTREACH PROGRAM IN BRAZIL

Andréa Oxley Da Rocha*, João Antônio Bonatto-Costa,
Júlia Pedron, Maria Paula Oliveira de Moraes
and Deivis De Campos

*Department of Basic Health Sciences, Federal University of Health Sciences
of Porto Alegre, Porto Alegre, Brazil*
*oxley@ufcspa.edu.br

ABSTRACT

In 2008, the Body Donation Program was established at the Federal University of Health Sciences of Porto Alegre (UFCSPA) due to an increasing demand for bodies for teaching anatomy. Pioneer in Brazil, the Program aims to inform the population about the possibility of donating one's body in life and its establishment allowed other interconnected activities to be developed by the Discipline of Human Anatomy, which are organized as an Anatomy Outreach Program. First among these activities, the Dissection Workshop is a 40-hour course that allows undergraduate students to enhance their knowledge of Anatomy and produce high quality prosections for gross anatomy classes. Once the Dissection Workshop ends, the best specimens are selected for annual temporary exhibition entitled Museum of Anatomy. The Museum is open to the community and organized by volunteer undergraduate students from preparing the rooms to acting as volunteer guides during visits. Lastly, an Ecumenical Ceremony to Honor the Body Donor is organized every year by first year medical students. In the presence of other students, professors and the donors' families, these students sing, play music and thank the families for their altruistic attitude, thus providing an emotional experience for everyone in attendance.

INTRODUCTION

Human bodies continue to be widely used for teaching anatomy in universities around the world. Despite the development of 3D software models that closely resemble human models, they are no substitute for the learning provided during the dissection of cadavers. Universities that once abandoned the use of human bodies for teaching purposes have re-introduced their use due to notable reduction in the quality of student learning and training (Rizzolo and Stewart, 2006). Thus, as has been shown in various studies (Aziz *et al.*, 2002), new technology for teaching anatomy should serve as a complement to the use of cadavers and not as a replacement. Moreover, proximity to the body plays an important role not only from a technical point of view, but also in the humanistic education of the students, who learn to respect human beings, particularly when they are in their most vulnerable state (Prakash *et al.*, 2007).

Voluntary body donation is used by many medical schools as the sole means of obtaining bodies for teaching anatomy in many countries, including Austria, France, Portugal, Germany, Romania (McHanwell *et al.*, 2008), Ireland (Cornwall *et al.*, 2012), China (Shang and Zhang, 2010; Chiu *et al.*, 2012), the United States (Garment *et al.*, 2007), New Zealand (McClea, 2008; Cornwall *et al.*, 2012), South Korea (Park *et al.*, 2011), India (Ajita and Singh, 2007; Ballala *et al.*, 2011), Japan (Kozai, 2007), the United Kingdom (Gangata *et al.*, 2010), South Africa (Kramer *et al.*, 2008; Cornwall *et al.*, 2012), Taiwan (Lin *et al.*, 2009), and Thailand (Winkelmann and Güldner, 2004; Prakash *et al.*, 2007). In each country, there is specific legislation for the regulation of the activity.

In Brazil, the situation is different. Most universities in the country still use unclaimed bodies for teaching anatomy, although each year the availability of such bodies decreases. In addition, most of the population is unaware of the possibility of voluntarily bequeathing their own body. Given this situation, in 2008, the Human Anatomy Department at the Federal University of Health Sciences in Porto Alegre (*Universidade Federal de Ciências da Saúde de Porto Alegre* — UFCSPA) introduced the Body Donation Program for Teaching and Research in Anatomy (Da Rocha *et al.*, 2013). In 2010, as part of that program, the first Ceremony in Honor of the Body Donors, a pioneering event in Brazil, was held.

Memorial ceremonies that express gratitude to body donors reflect a paradigm shift among medical schools with regard to the attitudes of the professionals undergoing training in medical programs, with greater emphasis "on humanity and compassion" (Jones *et al.*, 2014). This shift has led many anatomy programs to view the corpses used for the purpose of study as the "first patients" or the "teachers" or encouraging the students to understand that the donors entrusted their bodies in the belief that by doing so they would help improve academic training (Weeks, 1995; Ferguson *et al.*, 2006; Prakash *et al.*, 2007; Zhang *et al.*, 2008; Talarico, 2013; Jones *et al.*, 2014).

THE CEREMONY

Since 2010, at the end of each academic year, first-year students (freshmen) from the undergraduate courses and scholarship holders within the Human Anatomy Extension Program, as well as their teachers organize the Ecumenical Ceremony in Honor of Body Donors, as part of the Body Donation Program (BDP). This ceremony has become a tradition at the university and is keenly awaited by the students starting the course in gross anatomy.

The students participate in every step of the ceremony, from its planning and organization to its coordination and realization. Students who are finishing the course in gross anatomy that year are responsible for planning the ceremony, decorating the venue and welcoming the guests, as well as selecting and playing music for the ceremony and inviting a religious representative, based on a consensus among the students. For their part, the undergraduate students, scholarship holders and volunteers linked to the Anatomy Outreach Program, peer mentors along with the teachers, are responsible for the bureaucratic and logistical issues related to the ceremony's organization, including publicizing the event and sending invitations to University staff and relatives of the donors.

In an atmosphere of union and cooperation between students and teachers, the event begins with a song to welcome the guests sung by the students, who attend the ceremony dressed in their traditional white coats to facilitate their identification by the members of the donors' families. There follows, a series of short speeches of thanks and reflection given by the rector of the University, the coordinator of the Body Donation Program, and a representative of the first-year undergraduate students chosen by their peers. This is an opportunity for the students to freely express their feelings and thoughts regarding body donation and the role it plays in their professional, ethical and humanistic education as well as reflect on death and its significance.

Each year, a representative from a religious group has been invited by the students. Since 2010, when the ceremony was first performed, representatives of the Catholic Church, Spiritualist movement, Buddhism, the Lutheran Church and of the Seicho-No-ie philosophical movement have participated, illustrating the laical nature of the ceremony. Although the ceremony is not religious in nature, a word from a religious representative encourages reflection and provides a moment of comfort for the families of the donors present.

Then, the "Candles Ceremony" is held, in which students carrying candles that represent each of the bodies received by the BDP that year enter the room and another song is sung by the students. This is one of the most deeply emotional moments of the ceremony. After, photos of the donors (provided by their families) are projected onto a screen while their names are read aloud by the students. At the end, the students distribute flowers to the family members, along with a message of thanks for their selfless gesture and again sing popular songs with positive messages (Fig. 1). Every year, according to the students' decisions, the ceremony can be modified and new elements added. During the ceremony held in November 2015, the University choir was invited to appear at the end of the event, by the exit of the building, to surprise and raise the spirits of the visitors and students (Fig. 2).

After the end of the ceremony, when there is a moment of greater proximity between them, students have the opportunity to say goodbye to

Figure 1. (a) Flowers given to the relatives of the body donors. (b) "Candles Ceremony". (c) Students gathered around the family members. In the background, photos of the body donors are displayed on the screen.

Figure 2. (a) The University's choir performing to the family members of the body donors at the end of the ceremony. (b) Students and professors gathered at the end of the Memorial Ceremony. (c) First-year undergraduate students singing at the memorial ceremony.

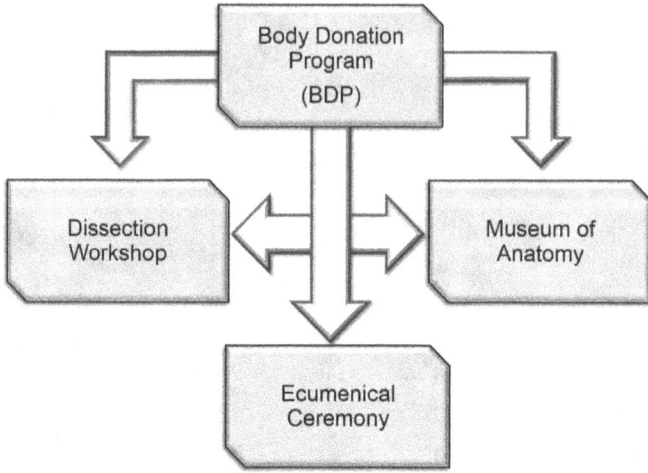

Figure 3. The interdependent relationship between projects in the Anatomy Outreach Program.

the members of the donors' families. The emotion of the students when they see the photos of the donors and realize the 'humanity' of the bodies they used for in their learning is evident, and translated into tears and smiles.

This tribute to the donors is one outcome of the BDP. And so, due to the success of the program in terms of the increase in the bodies obtained and, consequently, the greater availability of material for teaching gross anatomy, it has been possible to develop other two interrelated projects — the Dissection Workshop and the Anatomy Museum — together constituting the Anatomy Outreach Program, now well-established in the University (Fig. 3). Underpinning this set of activities is the belief that the extensive involvement of the students in the organization and execution of the projects is key to their growth and maintenance, which in turn contributes to the technical and ethical education of the students.

BODY DONATION PROGRAM

As currently occurs in other Brazilian universities, until 2007, most of the cadavers used for teaching at UFCSPA were unclaimed bodies from the

Forensic Medicine Institute of Rio Grande do Sul State, Brazil. As from 2008, the Body Donation Program for Teaching and Research in Anatomy was developed and approved by the Ethics in Research Committee of UFCSPA (no. 408/08). Its purpose being to inform the population, through publicity and awareness campaigns, of the possibility of voluntarily donating the body in life, and how to make such a donation. Therefore, this program uses documents that ensure the integrity and legality of the body donation process in life and after death.

This program is organized into two distinct areas. The first relates to the organization of campaigns designed to inform the general public of the possibility of body donation and the dissemination of information about the program to members of the academic community of UFCSPA. Newspaper articles and radio interviews are used to clarify the main questions from the general public regarding body donation, while folders emphasizing the importance of body donation and explaining how to proceed with such a donation are distributed mainly at events organized by the university. The second is linked to guidance for individuals who contact the BDP to make donations.

In addition to the necessary legal documents, at the time of the donation, the donor or family member, completes a form requesting identification and demographic data together with other relevant information, which is stored in a computerized database. The information on this database can be used to analyze donor profiles and thereby devise strategies for maintaining and disseminating the program.

THE DISSECTION WORKSHOP

With the good results obtained with the Body Donation Program, seen in the gradual increase in the number of bodies received through voluntary donation, the Dissection Workshop was created. The Workshop is a 40-hour extension course offered annually to undergraduate students from different courses in the area of health from the university who have already taken the Anatomy course. The aim is to encourage students to dissect by providing further knowledge in anatomy, while developing surgical skills, exercising teamwork and offering the possibility to produce scientific research. Activities in the form of practical lessons take place in the

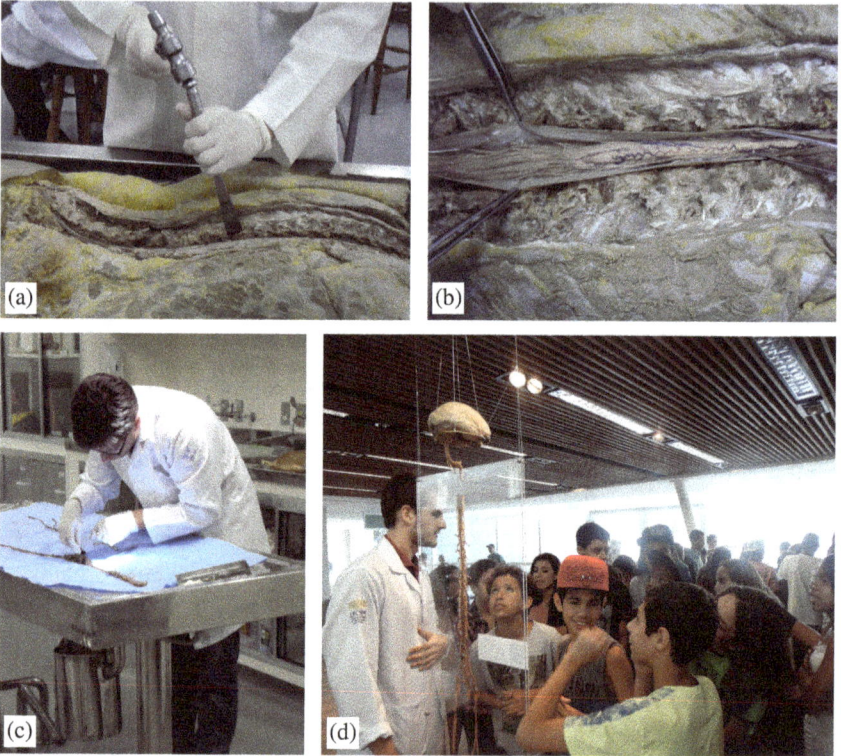

Figure 4. (a), (b), (c) Students prepare anatomical specimens (spinal cord) during the Dissection Workshop. (d) Anatomical specimen after dissection is concluded in exhibition at the Anatomy Museum as presented by a volunteer guide (in white coat).

anatomy laboratory, guided by teachers from the department and peer mentors. The materials produced during this course (prosections) are used for didactic purposes in practical classes of gross anatomy and particularly high quality specimens are included in the anatomy laboratory's collection, to be exhibited in the Anatomy Museum (Fig. 4).

THE ANATOMY MUSEUM

This is an annual event that has been held since 2008. It consists of a temporary exhibition of anatomical specimens produced by the students during the dissection workshop, by the teachers from the Human Anatomy department and from the laboratory's collection. In addition, a collection of replicas of famous works of art that show the close connection between anatomy and art are also displayed. The Museum also makes use of technological devices such as tablets with applications developed by the students themselves, where visitors can access information on the exhibited specimens and the human body systems.

Lasting 7 days, the Museum occupies spaces within the university designated for that purpose. All the work involved in organizing the physical space, as well as preparing the specimens for display is done by the volunteer students, scholarship holders linked to the project and teachers from the Anatomy department (Fig. 5). Also, due to the high demand, an

Figure 5. (a) Volunteer guide presenting replicas of artworks related to anatomy during the exhibition. (b), (c) Undergraduate students working in the assembly and organization of the Anatomy Museum.

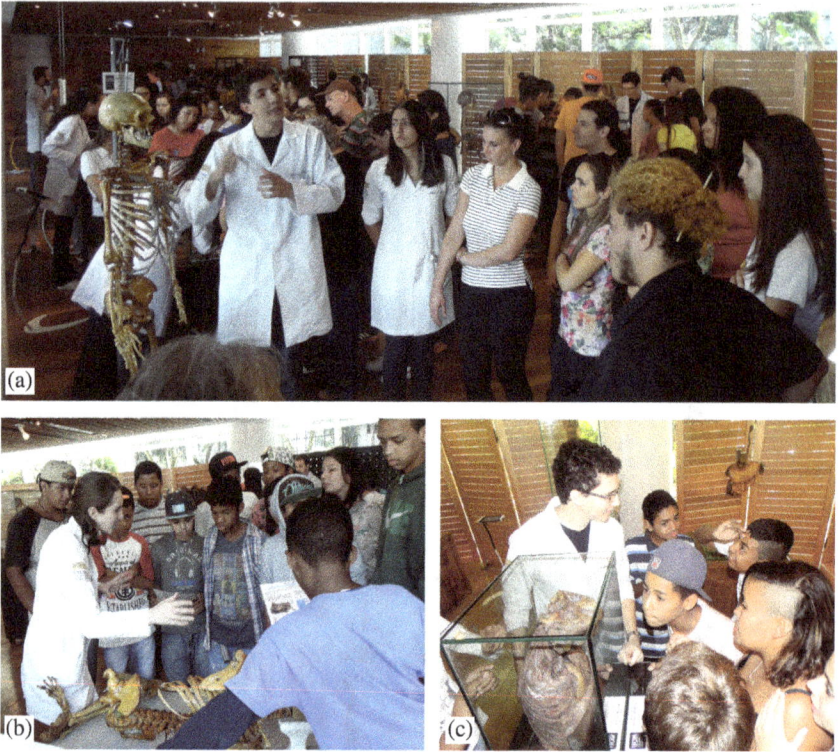

Figure 6.   (a), (b), (c) Volunteer guides explain the anatomical specimens to students from public schools at the Anatomy Museum.

open selection process is held for undergraduate students who want to volunteer as visitor guides during the exhibition. Entry to the exhibition is free and open to members of the internal academic community (students, staff and UFCSPA teachers) and the wider community, as well as public and private schools. All the visits take place in small groups, guided by undergraduate students (Fig. 6).

Over the years, the Museum has not only been successful in terms of the growth in the number of visitors, but also in their growing interest. Every year more and more students apply to act as guides, as well as compete for scholarships to participate in the project. The number of visitors has grown considerably since the first edition in 2008, when there were 170 visitors, to 2015, when there were 6,597 visitors during the seven days

Results of Exhibition

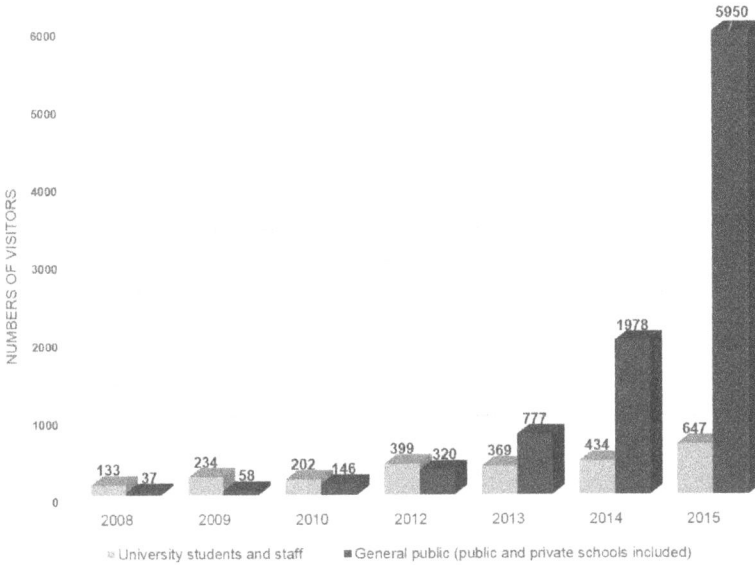

Figure 7. The Museum of Anatomy's evolution from its first to its seventh edition. The numbers above the columns represent the total of visitors that attended the exhibition in an specific year, from each of the two categories: visitors who reported themselves as part of the University faculty and staff and University students (light grey); and visitors who came from public and private schools, as well as other Universities and institutions, referred to as the general public (dark grey).

of the exhibition. Moreover, there has been a change in the composition of those attending. In 2008, only 22% came from outside the academic community, while in 2015, that figure was 90% of the total, indicating that the overall increase in visitors is due mainly to the increased interest of the community outside the university (Fig. 7).

These numbers, combined with the results from satisfaction surveys conducted among visitors and guides, suggest that factors such as the improved quality and quantity of the exhibits as well as the introduction of new elements has led to an increase in the numbers of visitors from the wider public and stimulated their curiosity about human anatomy. For the guides, the exhibition contributes to their academic training by enhancing their teaching skills and capacity for teamwork, as well as expanding their knowledge of anatomy.

Thus, the direct involvement of the students in the activities related to the Museum, i.e. the preparation of the anatomical specimens, assembly of the displays, and then, as guides, interacting directly with the public, who appreciate their work, induces the students' total commitment. Furthermore, this process increases the awareness of those students regarding the importance of body donation, their responsibility in relation to their own learning, and their pride in sharing academic knowledge with members of the community.

In addition, the Anatomy Museum has the effect of demystifying the use of the human body for anatomy teaching and research by showing the general population how the bodies donated to the University (BDP) are used.

DISCUSSION

In Brazil, most people believe medical schools only use unclaimed bodies for teaching and research. Moreover, they think that is not a problem because the bodies are abandoned. However, we are working to change this paradigm. Body donation programs are slowly expanding and informing the public of the possibility of voluntary donation in life.

In this context, the undergraduate students acting as agents disseminating this information and the memorial ceremonies in honor of the donors play an important role. The ceremony and the related activities lead the students to understand and reflect on the fact that volunteer donors provide their bodies altruistically, in the belief that, by doing so, they are contributing towards creating better trained professionals and that the students, therefore, should be committed to their learning. In addition, these ceremonies allow students to recognize and express their gratitude to the donors and their families for this gesture, which generates an increased sense of professional responsibility. Furthermore, this provides the students with the opportunity to articulate what they understand by life and death and, in doing so, reflect on their own mortality (Pawlina *et al.*, 2011).

Although this ceremony is clearly intended to express gratitude to the donors and their families, a religious aspect has always been included by the students. That might be due to the fact that although Brazil is a secular state, about 92% of the population claims to have some sort of religious belief (2010 census data). In some Eastern cultures, ceremonies in honor of body donors are essentially religious in nature (Subasighe and Jones, 2015). However, in a study of the memorial ceremonies held in anatomy programs in the United States (Jones *et al.*, 2014), the authors describe the use of religious verse, text, or prayer in only 6% of such ceremonies and suggest religious readings are more likely to be left out of the ceremony due to the heterogeneity of the participants.

Another interesting aspect concerns donor confidentiality because there are contrasting views regarding this matter. In some anatomy programs the identity of the donor is maintained confidential (Jones *et al.*, 2014). However, there is a tendency towards reinforcing the identity of the donors as individual human beings rather than as anonymous cadavers, in the hope of realigning student doctors' attitudes to their future patients (Talarico, 2013; Williams *et al.*, 2013; Jones *et al.*, 2014). In 2013, we started to mention the donors' names during the ceremony because we believe that a more personal nature of the ceremony, achieved by citing the donors' names, enhances the act of welcoming and expressing gratitude to the families of the donors, while providing them with an opportunity to remember and say farewell to their loved ones, since they have given up their traditions and family funeral rites by accepting the decision of the body donor. In addition, the large number of students present, and the respect shown towards the donors, transmitted through their words and gestures during the ceremony, demonstrate seriousness and reassure the families regarding the care taken with each donor's body.

Hence, the Outreach Program as a whole depends for its development on the involvement of the students, both as scholarship holders and volunteers involved in the projects, working to achieve their organization and implementation as well as participants in the activities and courses. Students at the end of the first year of the medical school, when they are taking the course in anatomy, participate in organizing the Ceremony. As from the second, third and fourth year, they can attend the dissection workshop and compete for scholarships or participate in the various projects on a voluntary basis.

CONCLUSION

Today the Ceremony functions as a link uniting all the activities in the Anatomy Outreach Program, while, at the same time, providing an outcome for these events. Through the Dissection Workshop and the Anatomy Museum, we provide students with greater technical and scientific skills and simultaneously heighten their awareness of issues related to the need and importance of body donation in life for the continuance of those activities. We also deal with ethical issues in an attempt to change paradigms, beginning from the first year of the course, through the Ceremony, raising awareness of their responsibility and dedication to their learning, assuming the commitment to return the knowledge received from the university to the public in the form of education and art, through the Anatomy Museum, and, with their enhanced training, provide better quality service to their future patients.

REFERENCES

Ajita R, Singh YI. 2007. Body donation and its relevance in anatomy learning–A review. *J Anat Soc India* 56:44–47.

Aziz MA, Mckenzie JC, Wilson JS, Cowie RJ, Ayeni SA, Dunn BK. 2002. The human cadaver in the age of biomedical informatics. *Anat Rec* (New Anat) 269:20–32.

Ballala K, Shetty A, Malpe SB. 2011. Knowledge, attitude, and practices regarding whole body donation among medical professionals in a hospital in India. *Anat Sci Educ* 4:142–150.

Chiu HY, Ng KS, Ma SK, Chan CH, Ng SW, Tipoe GL, Chan LK. 2012. Voices of donors: Case reports of body donation in Hong Kong. *Anat Sci Educ* 5:295–300.

Cornwall J, Perry GF, Louw G, Stringer MD. 2012. Who donates their body to science? An international, multicenter, prospective study. *Anat Sci Educ* 5:208–217.

Da Rocha AO, Tormes D, Lehmann N, Schwab R, Canto R. 2013. The body donation program at the Federal University of Health Sciences of Porto Alegre: a successful experience in Brazil. *Anat Sci Educ* 6:199–204.

Ferguson KJ, Iverson W, Pizzimenti M. 2006. Constructing stories of past lives: Cadaver as first patient: "Clinical summary of dissection" writing assignment for medical students. *Perm J* 12:89–92.

Gangata H, Ntaba P, Akol P, Louw G. 2010. The reliance on unclaimed cadavers for anatomical teaching by medical schools in Africa. *Anat Sci Educ* 3:174–83.

Garment A, Lederer S, Rogers N, Boult L. 2007. Let the dead teach the living: the rise of body bequeathal in 20th-century America. *Acad Med* 82(10):1000–5.

Instituto Brasileiro de Geografia e Estatística — IBGE. 2010. Censo Demográfico — 2010. Rio de Janeiro: IBGE. http://www.ibge.gov.br/estadosat/temas.php?sigla=ac&tema=censodemog2010_relig

Jones TW, Lachman N, Pawlina W. 2014. Honoring our donors: A survey of memorial ceremonies in United States anatomy programs. *Anat Sci Educ* 7:219–223.

Kozai T. 2007. History of collecting cadavers in Japan. *Kaibogaku Zasshi* 82:33–36.

Kramer B, Pather N, Ihunwo AO. 2008. Anatomy: spotlight on Africa. *Anat Sci Educ* 1:111–118.

Lin SC, Hsu J, Fan VY. 2009. "Silent virtuous teachers": Anatomical dissection in Taiwan. *BMJ* 339:19–26.

McClea K. 2008. The Bequest Programme at the University of Otago: Cadavers donated for clinical anatomy teaching. *N Z Med J* 121:72–78.

McHanwell S, Brenner E, Chirculescu ARM, Drukker J, van Mameren H, Mazzotti G, Pais D, Paulsen F, Plaisant O, Caillaud MM, Laforêt E, Riederer BM, Sañudo JR, Bueno-López JL, Doñate-Oliver F, Sprumont P, TeofilovskiParapid G, Moxham BJ. 2008. The legal and ethical framework governing body donation in Europe — A review of current practice and recommendations for good practice. *Eur J Anat* 12:1–24.

Pawlina W, Hammer RR, Strauss JD, Heath SG, Zhao KD, Sahota S, Regnier TD, Freshwater DR, Feeley MA. 2011. The hand that gives the rose. *Mayo Clin Proc* 86:139–44.

Prakash P, Rai R, D'Costa S, Jiji P, Singh G. 2007. Cadavers as teachers in medical education: knowledge is the ultimate gift of body donors. Singapore *Med Journal* 48:186.

Rizzolo LJ, Stewart WB. 2006. Should we continue teaching anatomy by dissection when...? *Anat Rec B* (New Anat) 289:215–218.

Shang X, Zhang M. 2010. Body and organ donation in Wuhan, China. *Lancet* 376:1033–1034.

Subasighe SK, Jones DG. 2015. Human body donation programs in Sri Lanka: Buddhist perspectives. *Anat Sci Educ* 8:484–489.

Talarico EF Jr. 2013. A change in paradigm: Giving back identity to donors in the anatomy laboratory. *Clin Anat* 26:161–172.

Weeks SE, Harris EE, Kinzey WG. 1995. Human gross anatomy: A crucial time to encourage respect and compassion in students. *Clin Anat* 8:69–79.

Winkelmann A, Güldner FH. 2004. Cadavers as teachers: The dissecting room experience in Thailand. *BMJ* 329:1455–1457.

Williams AD, Greenwald EE, Rhonda LS, DePace DM. 2013. Medical students' to reactions to anatomic dissection and the phenomenon of cadaver naming. *Anat Sci Educ* 7:169–180.

Zhang L, Wang Y, Xiao M, Han Q, Ding J. 2008. An ethical solution to the challenges in teaching anatomy with dissection in the Chinese culture. *Anat Sci Educ* 1:56–59.

11

A MOMENT OF HOPE: THANKSGIVING SERVICES FOR FAMILIES OF DONORS

Nalini Pather* and Ken Ashwell

School of Medical Sciences, UNSW Australia, Sydney, Australia
*n.pather@unsw.edu.au

ABSTRACT

At the University of New South Wales, Australia, a Thanksgiving Service for the families of donors was initiated in 2002. This ceremony has grown to include students performing musical tributes, readings, prayers of thanksgiving and the laying of flowers and refreshments. This chapter describes this ceremony and its value and impact on families, students and staff.

INTRODUCTION

Commemoration services acknowledging the gift of the donor have a long history, occur in various formats and often acknowledge the cultural norms and values of the immediate society (Tschernig and Pabst, 2001; McClea, 2008; Kooloos *et al.*, 2010; Riederer, 2016). In Australia, there is a growing trend to institute formal services to acknowledge donors. At the University of New South Wales (UNSW) Australia, this service has grown to include family members, students performing musical tributes, poetry readings, prayers of thanksgiving and the laying of flowers. The commemoration also allows the informal interaction at the close of the ceremony between students, staff and the families over donors over a light supper provided by the school.

The current form of the UNSW Thanksgiving Service commenced in 2002 by one of the authors (KA) as a way for the staff and students of the Department of Anatomy to provide a public expression of their gratitude to the families of donors. We had previously acknowledged donors within services held in School of Medical Sciences on All Souls' Day (usually celebrated on 1st or 2nd November). These early services led by our then Gross Anatomy Laboratory Manager, Dr. Adrian D'Mello, were usually held in the anatomy mortuary and attended by the gross anatomy academic and professional staff and a few interested students, but did not provide an opportunity for a more public recognition of the donors.

Our current service is modeled on the long tradition at the Department of Anatomy at University of Queensland (UQ), where public recognition in the form of a Thanksgiving Service has been in place for several decades. Some changes to the original format were necessary for UNSW. With permission from UQ, several of the original prayers were modified, and contributions from our own students were incorporated, including accompanying artwork for the Thanksgiving Service Booklet that is published for attendants. While in the UQ tradition, a candle is lit to honor each donor, at UNSW a group of students take turns to lay a flower to symbolically recognize and acknowledge each donor as their names are read out. A further symbolism that we have introduced is a decorative tree with autumn leaves bearing the names of each of the donors. Both the flowers and leaves are given to a representative of each family at the conclusion of the service as a token of remembrance of the family member.

THE FAMILIES OF THE DONORS

The families of donors understand and accept the wishes of their loved ones in donating their bodies to anatomy. Nevertheless, the long period of separation (up to 8 years in some cases) between the death of their family member and the return and interment or cremation of the remains may make it difficult for them to gain closure. We find that providing a public service within a year of the donor's passing provides the families with an opportunity to see the great good that has come from the donation and to hear how highly we regard their loved one's gift. Although we recognize the loss of individual family members, we dissuade family members from paying individual tributes to their loved ones at these services. Many family members speak of their deeply felt gratitude for the service and the way it has helped them appreciate why their loved one donated his or her body, and is summarized in this quote from one of the families:

> *On behalf of the family of Alice Gardner we would like to thank you for providing an opportunity for us to come together with other families and acknowledge our mother's life, generosity and vision through the donation of her body to Anatomy Schools. The way in which our mother was acknowledged allowed us to see how this donation was valued and valuable to students and faculty alike. This was a great encouragement to us.*

STUDENTS PARTICIPATING IN THE CEREMONY

An equally important benefit of the Service is the tangible focus it provides for our students to explore ethics and professionalism (Lachman and Pawlina, 2006; Bockers *et al.*, 2010). All our anatomy courses begin with a discussion of the importance of treating human remains with care and respect, but it can be difficult for new students to make the connection between a living person and the embalmed cadaver or prosections that they study. Preparing for, and participating in, such a service has an enormous impact on helping students realize the privileges that have been given to them to study another person in the most intimate of ways. This was expressed by one student as:

> *"it is most humbling (sic) to realize that we have studied, discussed and gotten to know another so intimately. For me personally, it made me realize that as I study to be a doctor, I have a responsibility to the least of humanity."*

Students participate in the Thanksgiving Service voluntarily. Literature acknowledges the importance of providing students with occasions for not only acknowledging donors, but where needed, for opportunities for closure when they have found learning from cadavers emotionally stressful (Giegerich, 2002). The act of students volunteering is in itself a way in which they personally pay tribute to the donor:

> *Volunteering has always been one of my personal interests. I chose to volunteer to help with the Ceremony as a gesture to show my appreciation for the families who have generously donated the body of their loved ones to the university for study and research. Furthermore, through participating in the Ceremony, I can interact with the attending families and show our gratitude on behalf of the entire student body.*

Speaking with the families of donors after the ceremony gives our students a very meaningful insight into the life and personality of the donor and creates a connection that stays with them throughout their careers. Our students' reflections after these services include:

> *One cannot succeed alone without the help of others. The road to succeeding in life's endeavors is often the result of many others' sacrifices and chances created as a result. Similarly, being able to study and pursue what I love is the greatest joy and I could not have gotten here without the many sacrifices made along the way by many others. My attitude towards learning is thus to do it for those who do not have the chance to. The thought of our body donors making the decision to allow their bodies to be studied to aid our learning fills me with gratitude because each and every one of them are victors in life and in death. Many of them would have been in a state of illness prior to death and in my personal opinion, this courageous act of body donation signifies a transcendence of death itself to create meaning and value for further generations to come. Opportunities do not arise by accident and to me, this opportunity to express my sincere gratitude to the body donors as well as offer solace, gratitude and reassurance to their families was definitely the best way I can help in my limited capacity. This is why I volunteered: to be able to offer comfort as one human being to another so as to show, through my actions, the softer and humanistic side of this profession to the public.*

One of the postgraduate students who participated in the 2015 ceremony highlighted the importance of the community aspect of the service:

"It is a community gathering — of families and friends of donors, with staff and students who really understand the value of the gift of body donation. The acknowledgement was carried out sincerely from hearts of all participants, which was undoubtedly the most moving part of the Ceremony."

STAFF PARTICIPATING AND ATTENDING IN THE CEREMONY

An advantage of commemorating services that is often underappreciated is that it acknowledges the nobility and dignity of the work of professional staff who spend most of the time with the donor in the processes of preparing and embalming for dissection, meticulously preparing prosections, dealing with the logistical issues related to storing, caring and eventually returning the remains of the donor to the family. One of our professional staff expressed the impact of the ceremony on her as follows:

My experiences at the thanksgiving ceremonies have furthered my understanding of the priceless bequest that the UNSW donors and their families make. Some of the family members are interested in talking to the Anatomy Laboratory staff. I appreciate the opportunity to chat with and thank the donor families in person. Meeting the families helps put a realistic perspective on the "human value" of our work. The study of anatomy benefits greatly from the ability to teach with real human specimens. I like having the opportunity to speak with them about this. I feel that the donor families may find some extra comfort in knowing that their family member is assisting future medical practitioners. Many of the people I have spoken to at the ceremonies have been very happy and thankful that we have been able to fulfill their family member's final wishes. Every family member has a unique story and it can be wonderful to hear the reasons behind the donors wanting to donate. I believe that the families are offered an amount of closure by being invited to and attending the Thanksgiving ceremony. Interacting and socializing with other donor families and the staff who currently care for the donor can be helpful and therapeutic for both sides.

These commemorating services play a significant role in formulating academic staff's philosophy not only to donors but also to the responsibility to educate the next generation of medical and health care practitioners. One of the authors (NP) participated in these services from early in her academic career, and attests to its formative value in instilling a deeper appreciation for those who have chosen to give their bodies to medical universities for the purposes of teaching and research. This impact early in her career has been formative in ensuring that the use of these resources is effectively and ethically used.

An anatomy academic who began participating in the services only recently expressed the impact of participation in this reflection:

Attending a remembrance service with donor family members allowed me to reflect on my own moral obligations as an academic in how I use donated bodies and interact with students as well as the general public. I felt acutely aware of my position of influence and obligation not only to the individual donors but also to their families. The service was a great reminder to me that donors and their families trust in me to do what is 'right' with their mother, father, sister, brother. As academics, our attitudes and beliefs, and behavior influences students' perceptions regarding death, donation and human value and meaning. Meeting family members was a powerful reminder that donors were real people who were loved by others, and who are still remembered by their friends and families. This service was important for me as celebration of lives of those generous people who believed that even after their death they could help others. This service made me realize the bond between the community and the university in remembering, respecting and appreciating donors. It helped me link my professional responsibilities with my personal feelings and moral beliefs.

THE UNIVERSITY CHAPLAIN

In the last few years, one of the chaplains of UNSW, Reverend Andrew Johnson, has assisted as the Master of Ceremony of the Thanksgiving Service. His reflection on the Thanksgiving Service and its value places it not only in the context of the medical school, but also as a sign of hope for the human goodness:

> *The Thanksgiving Ceremony represents a unique moment when our private lives and public commitments can be held together. As a University chaplain it is one of my great privileges to help people weave together these threads of family grief, love and memory with a shared hope that medical science and community care are worth striving for. At the same time as we are remembering the wonder and folly of our uncle or mother, we are also looking forward to the unnamed student who is yet to come. We most likely will never know them, but we place our hope in their future life and work as a gift from a previous one. The ceremony is often solemn and at times sorrowful. Yet it is also a strange moment of hope when a whole room of families, and all the stories they represent, stand together to say that we do not live only for ourselves. We believe in the value and worth of something more. I continue to be involved in the ceremony because it is a strange, sorrowful, determined moment of hope amidst a too cynical world.*

ACKNOWLEDGEMENTS

The authors would like to acknowledge and thank the donors, past, present and future, for their unique gift that enables the teaching of anatomy to more than 1000 students each year at UNSW. We also acknowledge all staff, students and families involved in organizing, participating and attending the Thanksgiving Service at UNSW. A special thanks to those who contributed reflections to this paper.

REFERENCES

Bockers A, Jerg-Bretzke L, Lamp C, Brinkmann A, Traue HC, Bockers TM. 2010. The gross anatomy course: An analysis of its importance. *Anat Sci Educ* 3:3–11.

Giegerich S. 2002. *Body of Knowledge: One Semester of Gross Anatomy, the Gateway to Becoming a Doctor*. New York, NY: Scribner.

Kooloos JGM, Bolt S, van der Straaten J, Ruiter DJ, 2010. An altar in honor of the anatomical gift. *Anat Sci Educ* 3:323–325.

Lachman N, Pawlina W. 2006. Integrating professionalism in early medical education: The theory and application of reflective practice in the anatomy curriculum. *Clin Anat* 19:456–60.

McClea K. 2008. The Bequest programme at the University of Otago: Cadavers donated for clinical anatomy teaching. *NZ Med J*. 121:7278.

Riederer BM. 2016. Body donations today and tomorrow: What is best practice and why. *Clin Anat* 29:11–18.

Tschernig T, Pabst R. 2001. Services of thanksgiving at the end of gross anatomy courses: A unique task for anatomists? *Anat Rec* 265:204–205.

12

ANATOMY COMMEMORATIONS
ON YOUTUBE: A REVIEW

Yousef AbouHashem, Benjamin T. Brown and Goran Štrkalj*

Department of Chiropractic, Macquarie University, Sydney, Australia
*goran.strkalj@mq.edu.au

"The law of human nature is to communicate more efficiently."

Kim Dotcom, *Letter to Hollywood*

ABSTRACT

The use of social media in anatomy education has increased in recent years. Among the other possible areas of application, social media can be seen as an instrument to facilitate the push to imbibe anatomy and anatomy education with humanistic values and make practices in these disciplines more humane. This paper reviewed the presentation of anatomy memorial ceremonies on YouTube, one of the most utilized forms of social media. The review, which included only English language uploads, revealed 66 relevant videos. These videos showed ceremonies (full or in part), reflections and news reports, and were posted by either individuals or institutions, most of which were from North America. The ceremonies that were featured varied markedly with regards to timing, participation, terminology, and format. YouTube seems to function as an easily accessible platform for sharing experiences and ideas on anatomy memorials, and importantly the ways in which anatomy practices can be made more humane. It is suggested that YouTube and social media in general could be utilized even more resourcefully as a medium for the exchange of information relating to best practice in anatomy education.

INTRODUCTION

Social media has been one of the key vehicles for the exchange of information and ideas in the last two decades. These media have enabled the efficient transfer of data from multiple sources to multiple receivers, opening a variety of possibilities for communicating, exchanging and creating knowledge (Pavlik and McIntosh, 2015). One of the most popular forms of social media has been YouTube, a web-platform that enables users to post, watch, share, rate and discuss videos. It has been shown that YouTube can be utilized in a number of different settings, including education (Sherer and Shea, 2011; Arnbjörnsson, 2014). YouTube has demonstrated value in enabling users to make social impact in a relatively simple and cost-effective manner. For example, the anti-bullying campaign — It Will Get Better Project — aimed at lesbian, gay, and bisexual teenagers, began as a single video-post on YouTube and has transformed into a far-reaching campaign involving a number of influential personalities (Hartlaub, 2010).

Social media in general, and YouTube in particular, have been utilized with an increasing frequency in anatomy learning and teaching. Several recent studies have commented on the potential usefulness and ease of utilization of YouTube in anatomy education (Jaffar, 2012; Raikos and Waidyasekara, 2014). This is not surprising as one of the main developments in modern anatomy education has been the increasing use of technology (Chan and Pawlina, 2015).

At the same time anatomy education has undergone a cultural/ paradigm shift toward a more humane attitude, specifically in relation to the acquisition and treatment of cadavers for study in anatomical laboratories (Dyer and Thorndike, 2000; Talarico, 2013; Štrkalj, 2014). The observed shift seems to have been accelerated by the popularity and increased utilization of social media (Zhang *et al.*, 2014).

One of the important elements of this new, complex and multifaceted approach (Talarico, 2013; Hildebrandt, 2016a) in humanizing anatomy are commemorations (memorial ceremonies) carried out to honor the body donors. These ceremonies forge a relationship of trust and understanding between anatomists and their students on the one side, and body donors, their families, and society as a whole on the other (Bolt, 2012). In this role, social media are utilized in a variety of ways. For example, websites can be created which enable students and family members of body donors to interact, pay respect and express their gratitude (Zhang *et al.*, 2014). YouTube has featured prominently among the various forms of social media utilized in sharing information about memorial services in anatomy programs. The aim of this chapter was to review YouTube videos dealing with anatomy commemoration in anatomy programs and perform an analysis of the format and content.

METHODS

A search was performed using the following key terms: "anatomy", "cadaver", "gratitude", "thanksgiving", "memorial", "commemoration", "appreciation", "service", "ceremony" and "tribute". The search was carried out between the 26th of October 2015 and the 10th November 2015, using (A) the search engine on the YouTube site and (B) the advanced search function on the Google webpage. The same keywords that were used in the YouTube search were used to identify relevant content on Google, with the addition of appropriate search operators. To be included in the review a video had to include footage of either a complete or partial anatomy memorial service. In addition, highlights, tributes, news stories and reflections on anatomy memorial services were also reviewed. Videos that were clearly defined as an anatomy memorial service in the title or descriptor were included. To be eligible, videos had to be published in English or have appropriate subtitles annotating the events in the video. Videos in which the reviewers could not determine the exact nature of the memorial service were excluded, as were videos describing memorial services for organ donors.

Two reviewers independently carried out the search and screened the results. Irrelevant videos were omitted based on the title and/or descriptor where applicable. Relevant videos were viewed in full and analyzed by two reviewers. Salient details regarding the format and content of the videos was recorded using Excel. The reviewers then compared their individual results and compiled a master spreadsheet, with any disagreement settled by consensus.

RESULTS

The YouTube search resulted in the retrieval of 705 videos, while the Google search returned 529 relevant videos. After the removal of duplicate content and the application of the selection criteria, there were 66 videos that were suitable for inclusion in the review (Fig. 1). Much of the Google search results included the content found on YouTube, however one additional video from the website "Vimeo.com" was retrieved using this strategy. This video was not included in the final analysis.

The majority (62%) of videos presented were either complete or partial ceremonies, or highlights of a memorial event. Also featured were news reports, tributes (videos created for the viewing during the memorial service) and videos portraying interviews with individuals relating to the ceremonies and/or reflections about body donation and commemorations.

The duration of the videos ranged from 5 seconds to 94 minutes. The average length of the videos showing partial ceremonies or highlights was 5 minutes, while the average length of the full ceremony videos was 58 minutes. The video and audio production quality varied considerably. Both video and audio quality were scored on a 3-point scale: 1 = poor, 2 = moderate, 3 = good. The video and audio production quality scores were: 21 poor, 19 moderate and 26 good, and 24 poor, 20 moderate and 22 good, respectively. Collectively, the videos in the sample had been viewed 37,936 times, with the average number of views for each video being 574.

The video descriptors provided by the uploader varied from having no description through to very detailed descriptions. The main theme of the descriptors was either a description of the event or a description of the video. Other themes included song lyrics, lines or a stanza from a poem, the names of the performers, and notes on the video quality or the upload device. The word count of the descriptors ranged from zero to 544, with an average of 54 words. The descriptors that incorporated lines or stanza from a poem, or song lyrics tended to be longer in length.

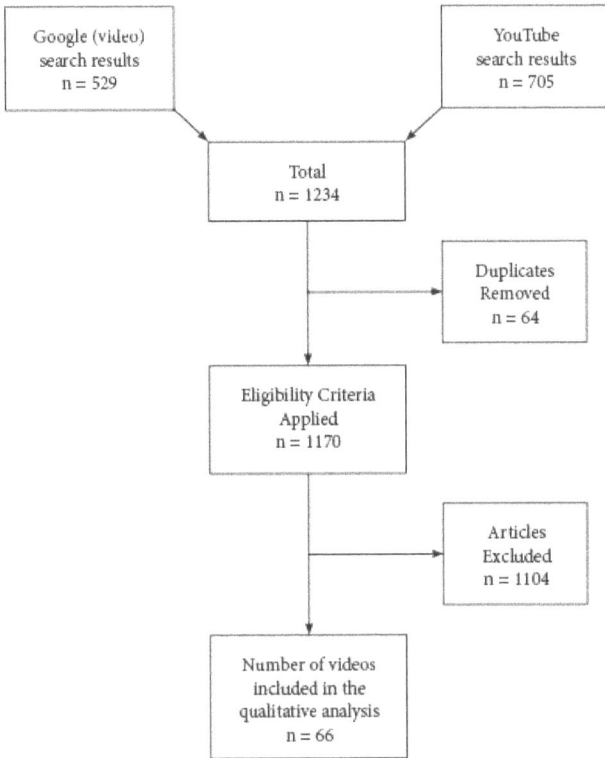

Figure 1.   Video search strategy.

When uploading content on YouTube, users can select any one of 15 options by which to categorize the content of their video. These options enable uploaders to more accurately target their desired audience. With reference to the videos in the sample, the common categories are listed in Fig. 2.

Videos were produced mainly by individuals (53%) and educational institutions (37%). Some videos were created by news agencies (9%). The vast majority (80%) of the videos originated from the United States, particularly the east coast. Other countries represented included Canada and Taiwan. There were eight videos in which the origin could not be determined. The earliest upload was in 2006, and since then, the number of uploads has been increasing, peaking in 2012 and 2013 (See Fig. 3 for a complete breakdown regarding the year of upload). The majority of videos were produced by educational institutions within a faculty of

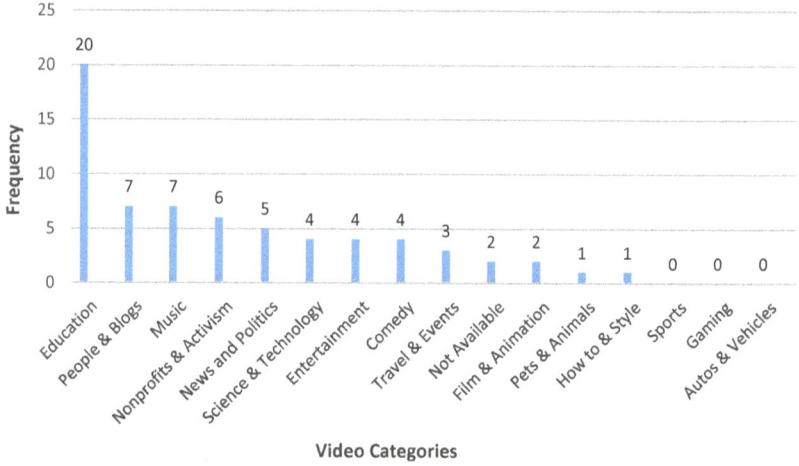

Figure 2.   Video categorization.

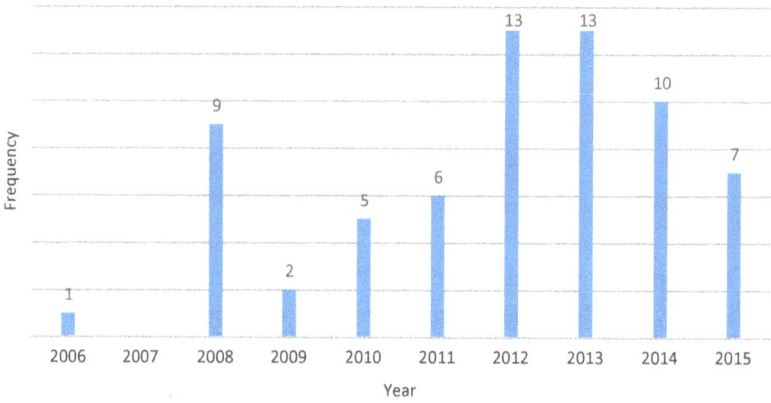

Figure 3.   Year of upload.

medicine. Students represented in the videos were those studying towards degrees in medicine, dentistry, osteopathy, physical therapy, mortuary studies as well as those studying to become physician assistants.

With reference to terminology, the majority of the videos showing commemoration, used the prefix "anatomy", "cadaver" or "donor" and referred to the event as either a "memorial", "memorial service", "ceremony" or "thanksgiving". The most common title string was "cadaver memorial service". One video used the phrase "appreciation ceremony". The deceased were most commonly referred to as "cadavers" or "donors". Other less commonly used terms were "gift", "bequest", "teacher/silent teacher" and "mentor/silent mentor".

Timing of the memorial ceremony was very similar across the sample with most ceremonies taking place at the end of the teaching period. Some schools held ceremonial events both at the beginning and the end of the anatomy course, while others carried out the ceremony at the time of the funeral service for the body donor. It was difficult for reviewers to ascertain exactly when the event took place in approximately half of the videos. In terms of location, the ceremonies were most commonly held in an auditorium, lecture theatre or concert hall. In many cases, the location was not specified in the description and it was difficult for reviewers to discern the exact type of venue from the footage. A small percentage of the ceremonies were held outdoors on university grounds, or a suitable space adjacent to a cemetery or a dedicated university memorial space.

The vast majority of ceremonies included staff and students from the respective school and friends and family members of the body donor. Reviewers were unable to determine the attendees in a large proportion of the videos as the guests were not mentioned or described in the footage. In a small percentage (9%) of videos, the ceremony was restricted to staff and students of the school. Anonymity of the donor was clearly preserved in 15% of the videos. In 15% of videos the donors first and last names were revealed either at the commencement or completion of the program. Reviewers were unable to determine if anonymity was preserved in the remaining 70% of videos.

There were 15% of videos that could be classified as secular based on the content of the ceremony. Some of these videos were described as interfaith ceremonies in the video description. A small percentage (12%) of videos contained prayers and could be classified as non-secular. The presence of a religious figure, e.g. priest, did not necessarily translate into the ceremony being representative of a particular religious denomination. Reviewers were unable to determine if the ceremony was secular in the majority of videos (71%).

Table 1. Expressions of gratitude.

Expression of Gratitude	Number of Videos	Percentage (%)
Speech	44	67
Instrumentals	21	32
Singing with accompaniment	25	37
Singing	16	24
Slide Show/Memorial Board/Video	15	22
Candles	13	19
Poetry	8	12
Reception	8	12
Moments of silence	7	10
Gifts to family members	6	9
Dance	4	6
Tour	2	3
Releasing of doves	1	2

There were a variety of ways in which gratitude/appreciation was expressed by attendees of the memorials for the deceased. Speeches by students, faculty and/or family members were common place in a large percentage (67%) of the videos. Other popular forms of appreciation were singing with accompaniment (37%) or instrumentals (32%). The forms of gratitude/appreciation that were expressed in the sample are displayed in Table 1. Many of the full ceremonies contained a number of different expressions whereas videos depicting partial ceremonies were more likely to be of single expressions e.g. a singing performance.

DISCUSSION

Recent historical accounts of anatomy strongly emphasized the dark side of the discipline which has been marked by abuse and ethically dubious practices in the acquisition of cadavers for dissection, and/or their treatment while in laboratories (MacDonald, 2005; Richardson 2000). Anatomists themselves have also brought this side of anatomy under historical and ethical scrutiny (Jones and Whitaker, 2009; Brenner and Pais, 2014; Hildebrandt, 2016b). Consequently, anatomy and the way it is taught have become, albeit not to the same degree in all places, more humane and compassionate, particularly in relation to bodies that have been donated to science. Honoring the body donor and establishing communication with donors, their family and society through memorial ceremonies has been one of the key features of this new approach to anatomy and anatomy education.

This review suggests that YouTube, as one of the commonly utilized social media platforms, has been increasingly employed to communicate information about the anatomy memorials. Furthermore, some of the videos posted on YouTube were produced specifically to be integrated in the commemorations, becoming part of the actual content of the memorial service. Videos also serve as a medium for reflection, which can be shared and discussed by a wide, global audience of individuals who share an interest in ethical and philosophical issues related to anatomy — primarily anatomists and their students, donors, and families, but also the general public.

Not surprisingly, the results of this review of YouTube videos were similar to those of a recent survey of memorial ceremonies carried out among the United States anatomy programs (Jones *et al.*, 2014). Both the review and the survey showed a terminological diversity with relation to the ceremonies and the bodies. This terminological heterogeneity is further accentuated in this review by the finding that there was wide variety of categories used to label the videos on the YouTube. This is to be expected bearing in mind that anatomy memorials are relatively novel phenomena. The terminological differences and resultant potential for confusion, highlight the need to discuss and critically examine the issues surrounding anatomy memorials and to improve the exchange of experiences between individuals and institutions.

Similarly, both the present review and the survey performed by Jones *et al.* (2014) revealed significant variation in the content and the format of the ceremonies. Both pieces of research highlight great variation both in the time of the academic year that the ceremonies were performed, and in the makeup of the attendees; they took a variety of forms, some being secular and others being conducted as religious (mainly non-denominational) events. They included a number of activities such as speeches, musical performances and poetry recitals as well as the use of ceremonial objects, such as candles. The ceremonies were performed in anatomy or other teaching facilities or at the places specially created to pay tribute to the donors.

There are some contentious issues relating to the anatomy ceremonies that were noted in this review. While there seems to be a consensus among anatomists as well as medical and science educators that performing memorial services is of great benefit (Jones, 2016), and advisable to all anatomy departments, some issues concerning the content of these ceremonies are still debated (Williams *et al.*, 2014). One of these issues is whether the identity of the donor should be revealed or kept anonymous. Both this YouTube review and the Jones *et al.* (2014) survey of the anatomy programs showed that in some ceremonies the identity of the donors was revealed, and their lives were celebrated at the individual level, while in others, as a sign of respect and right to privacy, donors were kept anonymous. The proportion of programs utilizing these two approaches was equal (47% anonymous and 47% not anonymous) in both Jones *et al.* (2014) study and this review (15%). Future research and debate, possibly also facilitated through social media, will shed more light on the issue of donor anonymity, particularly as applied within different cultural, political and legal contexts.

More than anything else, however, this review seems to show the great potential to incorporate YouTube into the process of disseminating humanistic values in anatomy, including the promotion of memorial services. Bearing in mind that the phenomena of donor commemorations in anatomy programs are relatively new, the ability to efficiently exchange ideas and facilitate social engagement (in which social media and YouTube can play an important role) seem to be particularly valuable (Tschernig and Pabst, 2001). A large majority of the videos reviewed in this paper originate from the North American universities (disproportionately high in relation to the number of countries in which English is the official, or one of the official, languages), institutions with relatively long and established tradition of memorial services in anatomy programs (Hildebrandt, 2010; Jones *et al.*, 2014). These experiences seem to be valuable to share, particularly with colleagues from countries in which such practices have still not been applied (Gangata *et al.*, 2010; Champney, 2011; Anyanwu, 2011).

The process of exchanging ideas relating to death and dying is likely to be complex as these concepts vary considerably across cultures. Indeed, just like the body donation programs, memorial ceremonies are, to a significant degree, shaped by the cultural values of the society within which they originate (Winkelmann and Güldner, 2004; Park *et al.*, 2011; Alexander *et al.*, 2014; Subasinghe and Jones, 2015). At the same time, however, approaches that have originated within certain cultural matrices can prove to be of great value outside of the cultural circle within which they originated (Bohl *et al.*, 2011).

Reaching out and facilitating change in regions in which humane and humanistic approach to anatomy and anatomy education has not developed to its full potential are tasks for the international anatomical associations and, indeed, each individual anatomist (Champney, 2011; Štrkalj, 2014; Jones, 2016; Riederer, 2016). It is therefore of great importance that good practices are shared with wider academic and even non-academic audiences. Because of the easy access and relatively low cost, YouTube provides an efficient vehicle for sharing information, which is illustrated by the North American experience with anatomy ceremonies. Indeed, YouTube has a much wider audience (both academic and non-academic) compared to a published manuscript or a commercial media piece and therefore greater potential impact.

Historians note that nineteenth century anatomy was secretive and purposely hidden from the public scrutiny (Richardson, 2000; Macdonald, 2005). In contrast, modern anatomy is marked by a culture of openness and transparency (Štrkalj, 2014). Therefore it may be argued that anatomy departments should be encouraged to strategically use social media, such as YouTube, even more broadly in communicating their experiences with strategies aimed at humanizing anatomy. Subscribing to the best practice in the acquisition and treatment of cadavers in anatomy laboratories and further developing these practices is one of the imperatives of modern anatomy and particularly anatomy education. As a tool, social media could play an even more prominent role in these processes.

CONCLUSION

YouTube has been utilized by individuals and institutions to communicate and share information about memorial ceremonies in anatomy programs.

It appears that YouTube and social media in general could play an even greater role making a social impact in promoting ethical and humane approaches to anatomy education.

REFERENCES

Alexander M, Marten M, Stewart E, Serafin S, Štrkalj G. 2014. Attitudes of Australian chiropractic students toward whole body donation: A cross-sectional study. *Anat Sci Educ* 7:117–123.

Anyanwu GE, Udemezue OO, Obikili EN. 2011. Dark age of sourcing cadavers in developing countries: A Nigerian survey. *Clin Anat* 24:831–836.

Arnbjörnsson E. 2014. The use of social media in medical education: A literature review. *Creative Educ* 5:2057–2061.

Bohl M, Bosch P, Hildebrandt S. 2011. Medical students; perceptions of the body donor as a "First Patient" or a "Teacher: A pilot study. *Anat Sci Educ* 4:208–213.

Bolt S. 2012. Dead bodies matter. *Med Anthropol Q* 26:613–634.

Brenner E, Pais D. 2014. The philosophy and ethics of anatomy teaching. *Eur J Anat* 18:353–360.

Chan LK, Pawlina W. (eds.) 2015. *Teaching Anatomy: A Practical Guide*. New York: Springer.

Dyer GSM, Thorndike MEL. 2000. *Quidne mortui vivos docent*? The evolving purpose of human dissection in medical education. *Acad Med* 75:969–979.

Champney TH. 2011. A proposal for a policy on the ethical care and use of cadavers and their tissues. *Anat Sci Educ* 4:49–52.

Gangata H, Ntaba P, Akol P, Louw G. 2010. The reliance on unclaimed cadavers for anatomical teaching by medical schools in Africa. *Anat Sci Educ* 3:174–183.

Hartlaub P. 2010. Dan Savage overwhelmed by gay outreach's response. *San Francisco Chronicle*, 8 October.

Hildebrandt S. 2010. Lessons to be learned from the history of anatomical teaching in the United States: The example of the University of Michigan. *Anat Sci Educ* 3:202–212.

Hildebrandt S. 2016a. Thoughts on practical core elements of an ethical anatomical education. *Clin Anat* 29:37–45.

Hildebrandt S. 2016b. *The Anatomy of Murder: Ethical Transgressions and Anatomical Science During the Third Reich*. New York: Berghahn Books.

Jaffar AA. 2012. YouTube: An emerging tool in anatomy education. *Anat Sci Educ* 5:158–164.

Jones DG. 2016. Searching for good practice recommendations on body donation across different cultures. *Clin Anat* 29:55–59.

Jones DG, Whitaker MI. 2009. *Speaking of the Dead: The Human Body in Biology and Medicine*. Farnham and Burlington: Ashgate.

Jones TW, Lachman N, Pawlina W. 2014. Honoring our donors: A Survey of memorial ceremonies in United States anatomy programs. *Anat Sci Educ* 7:219–223.

MacDonald H. 2005. *Human Remains: Episodes in Human Dissection*. Carlton: Melbourne University Press.

Tschernig T, Pabst R. 2001. Services of thanksgiving at the end of gross anatomy courses: A unique task for anatomists? *Anat Rec* 265:204–205.

Park J-T, Jang Y, Park MS, Pae C, Park J, Hu K-S, Park J-S, Han S-H, Koh K-S, Kim H-J. 2011. The trend of body donation for education based on Korean social and religious culture. *Anat Sci Educ* 4:33–38.

Pavlik JV, McIntosh S. 2015. *Converging Media: A New Introduction to Mass Communication*. 4th Edition. New York: Oxford University Press.

Raikos A, Waidyasekara P. 2014. How useful is YouTube in learning heart anatomy? *Anat Sci Educ* 7:12–18.

Richardson R. 2000. *Death, Dissection and the Destitute*. Chicago: The University of Chicago Press.

Riederer B. 2016. Body donations today and tomorrow: what is best practice and why? *Clin Anat* 29:11–18.

Sherer P, Shea T. 2011. Using online video to support student learning and engagement. *Coll Teach* 59:56–59.

Subasinghe K, Jones DG. 2015. Human body donation programs in Sri Lanka: Buddhist perspectives. *Anat Sci Educ* 8:484–489.

Štrkalj G. 2014. The emergence of humanistic anatomy. *Med Teach* 36:912–913.

Talarico EF. 2013. A change in paradigm: Giving identity to donors in the anatomy laboratory. *Clin Anat* 26:161–172.

Williams AD, Greenwald EE, Soricelli RL, DePace DM. 2014. Medical students' reactions to anatomic dissection and the phenomenon of cadaver naming. *Anat Sci Educ* 7:169–180.

Winkelmann A, Güldner FH. 2004. Cadavers as teachers: the dissecting room experience in Thailand. *Brit Med J* 329:1455–1457.

Zhang L, Xiao M, Gu M, Zhang Y, Jin J, Ding J. 2014. An overview of the roles and responsibilities of Chinese medical colleges in body donation programs. *Anat Sci Educ* 7:312–320.

13

CONCLUDING REMARKS:
WHERE TO FROM HERE?

Nalini Pather* and Goran Štrkalj†

*School of Medical Sciences, UNSW Australia, Sydney, Australia
†Department of Chiropractic, Macquarie University, Sydney, Australia

The emerging practice of commemorating donors may well represent a shift in anatomy education with far-reaching ramifications that include instilling a deeper appreciation for the need to embed humanistic values in medical and science students, adding dignity to the work of staff working with cadavers, and changing attitudes towards donors and donation.

In spite of increasing digital technology for teaching, cadaveric study remains an essential part of learning in medicine and in allied health professional programs. As the numbers of medical schools and medical students increase globally, the demand for cadavers continues to increase. The challenge for medical schools to acquire sufficient cadavers for teaching purposes is often difficult as the number of persons who choose to donate remains relatively low. Previous studies on organ and cadaver donation have identified mistrust of medical schools and hospitals, cultural attitudes and religion/spirituality as barriers to donation (Boulware *et al.*, 2002; Bolt, 2012). As shown by the chapters in this book, commemorating donors can play a significant and multi-faceted role in addressing these issues by establishing trust between members of academic institutions (students, teachers, administrations) and the local community in which the donors and their families form the essential nexus.

An interesting aspect of this discussion is attitudes to body donation. Arising from studies analyzing generational shifts in attitudes such as Twenge *et al.* (2012), is a growing concern that millennials, born in the 1980s and 1990s, are the least likely of all generations to donate. In order to preempt addressing this trend, we need to understand more about this generations' attitude towards donation, humane practices, health and benevolence. Early reports indicate that this generation responds well to story-telling through social engagement and media (Zhang *et al.*, 2014). As we step into the future of medical education we need to learn how to tap into the millennial culture of sharing, social media and emotional engagement so that we can foster a humane and humanistic approach to anatomy.

In order to ensure that the trend towards a humanistic approach to medical education continues, we need to be informed by robust research and wider communication of practices. Improving communication and trust between stakeholders, and developing programs targeting population attitudes towards donations are essential elements to addressing this problem.

REFERENCES

Bolt S. 2012. Dead bodies matter. *Med Anthropol Q* 26:613–634.

Boulware LE, Ratner LE, Cooper LA, Sosa JA, LaVeist TA, Powe NR. 2002. Understanding disparities in donor behavior: race and gender differences in willingness to donate blood and cadaveric organs. *Med Care* 40:85–95.

Twenge JM, Campbell WK, Freeman EC. 2012. Generational differences in young adults' life goals, concern for others, and civic orientation, 1966–2009. *J Pers Soc Psychol* 102:1045–1062.

Zhang L, Xiao M, Gu M, Zhang Y, Jin J, Ding J. 2014. An overview of the roles and responsibilities of Chinese medical colleges in body donation programs. *Anat Sci Educ* 7:312–320.

INDEX